D0238112

Site Surveying and Levelling

Second Edition

John Clancy

Dip. Arch. (Manchester), Dip. T. & R.P. (Leeds), A.R.I.B.A.

BUTTERWORTH
HEINEMANN

OXFORD AMSTERDAM BOSTON LONDON NEW YORK PARIS
SAN DIEGO SAN FRANCISCO SINGAPORE SYDNEY TOKYO

Butterworth-Heinemann
An imprint of Elsevier Science
Linacre House, Jordan Hill, Oxford OX2 8DP
225 Wildwood Avenue, Woburn MA 01801-2041

First published 1981
Second edition 1991
Reprinted 1993, 1995, 1997, 1999, 2001
Transferred to digital printing 2002

British Library Cataloguing in Publication Data
Clancy, John 1936–
    Site surveying and levelling – 2nd ed.
    1. Construction. Sites, levelling and surveying
    I. Title
    624

ISDN 0 340 50547 8

Typeset by Keyword, Wallington, Surrey
Printed and bound in Great Britain by Antony Rowe Ltd, Eastbourne

Contents

Preface to 2nd Edition

Since the publication of the first edition surveying, like most things, has developed in some areas but not in others and it is with this in mind that alterations and additions to the book have been carried out. Cognisance has also been paid to the comments of reviewers, student users and the experience of both colleagues and myself when using the work as a standard teaching text.

The basic philosophy of the book has not changed, in that my intention is still to deal with the material sufficient to ensure that the student will quickly be of use in the office or on site. For those intending to pursue higher studies I have added a new chapter relating to 'Modern Developments' covering the more basic surveying equipment and techniques now being used in the electronic age in which we find ourselves working. Again, the text is *an introduction* on which higher studies and work experience can be based rather than a definitive work. I hope that those students not intending to take higher studies will read the chapter out of interest.

In addition, new text has been introduced to cover a wider range of topics including Ordnance Survey; calculation of areas; computation of true horizontal length; measurement of vertical angles; Code of Measuring Practice; curve ranging and calculations of volumes for earthworks, whilst parts of the original text have been re-written and extended.

I hope, now, that most of the shortcomings inherent in the original work have been overcome giving a more useful, all-round text, not only to the beginner but also to those who, having a general interest in surveying, may not have practised the art for some time.

John Clancy
1991

Preface to 1st Edition

It has always seemed to me that most textbooks on surveying have set out to be 'definitive works' on the subject. This book has no such pretensions.

It is certainly a fact that all students attending classes and lectures on the subject of surveying do not intend to qualify as surveyors. It is also probably true that the majority of students do not become interested in practical subjects until they have left the learning situation.

Bearing the above in mind, this book aims to give a student an understanding of the basic principles of surveying. I have tried to deal with the subject in what I consider to be a logical and practical sequence, attempting always to make the book readable and interesting.

My intention is not, therefore, to cover the whole subject in one volume but to deal with sufficient material to make sure the student will quickly be of use to an office or site organisation.

The study of any subject should not be a chore – if it is, turn to something else – and hopefully the presentation of this book will turn surveying into a subject of interest from the outset. By sustaining this initial interest, the basics will be learned more thoroughly, thereby making later more advanced studies easier and – even more important – enjoyable.

John Clancy
1980

Acknowledgements

Again I must firstly acknowledge the unstinting assistance of my wife who, as with the first edition, typed, constructively criticised the text and generally ensured that deadlines were met.

Secondly, I would like to thank lecturing colleagues who have not only used the text but also suggested improvements – some of which have been incorporated – or provided illustrative material. I refer particularly to John Dunn of Central Manchester College of Technology and John Muskett of Bolton Institute of Higher Education. I must also record my appreciation for the help given to me by Mr. Bernard Lewis (Clarkson Group) for providing material for the first edition which has been re-used in the second edition and Mr. Stephen Blaikie of SOKKISHA for the use of photographic material in Chapter 10. I would also like to thank Mr. Ross Dallas of the photogrammetric unit, University of York, for his help with the material for Fig. 10.14.

The following organisations have also kindly given permission to reproduce photographs or illustrations: John Mowlem & Co. PLC (page viii); The French Government Tourist Office (page xiii, top); The Mansell Collection Ltd. (page xiii, bottom); Sotherby & Co. Ltd., for the end-piece to Chapter 1; A. Clarkson & Co. Ltd., (Figs 2.3, 2.13, 6.3, 6.5, 8.3, 8.4, 9.1 (board, tripod and micromptic alidade) and 9.2); Francis Barker and Son Ltd. (Figs 2.12, 2.16, 6.10 and 9.1 (alidade rule, trough compass and Indian-pattern clinometer) and Clyde Surveys (1990) Ltd, the successors to Fairey Surveys Ltd. (Fig. 10.13). The planimeter in Fig. 3.22(a) was taken from *Elementary Surveying* by John Malcolm and is reproduced by kind permission of © Unwin Hyman Ltd.

The examples of Ordnance Survey maps in Chapter 1 were reproduced from 1990 Ordnance Survey 1:25000 Ipswich, sheet TM04/14; 1: 10000 Cornwall – Devon, sheet SX47SW; 1:2500 Torquay, sheet SX9264 and 1:1250 Dunblane, sheet NS789NE with the permission of the Controller of Her Majesty's Stationery Office © Crown copyright.

A modern feat of surveying: the National Westminster tower in London is 183 m high but no more than 5 mm out of plumb, despite difficult site and ground conditions.

Introduction – what is surveying?

If the student consults a good dictionary, he will probably find a general definition of the word 'survey' like the following:

survey (*v.t.*) To inspect or take a view of; to view as from a high place; to view with scrutinising eye; to examine; to examine with reference to condition, situation, or value; to inspect for a purpose; to determine the boundaries, extent, position, natural features, etc. of, as of any portion of the earth's surface by means of linear and angular measurements and the application of geometry and trigonometry (and by photography from the air).

survey (*n.*) A general view; a look at or over; a close examination or inspection to ascertain condition, quantity, quality, etc.; the determination of dimensions and other topical particulars of any part of the earth's surface; the map, plan, or account drawn up of such particulars.

Types of surveyor

From the above definitions, the student can appreciate that there are many aspects covered by the general term 'surveying', and that, of necessity, the profession as a whole is made up of many different types of surveyor.

a) The general-practice surveyor

is one whose basic expertise is *valuation* in connection with purchase, sale, letting, investment, mortgage, rating, insurance, compensation, or taxation. He may also be involved in *estate agency*, concerned with negotiations for sale or purchase, leasing or auction of property, or *estate management*, concerned with the management and maintenance of residential, commercial, and industrial property on behalf of both landlord and tenant. Some surveyors within this category will specialise in the valuation and auction of furniture and works of art, plant, and machinery.

b) The building surveyor

offers a specialist service on all matters relating to construction, along with assessment of rebuilding costs for insurance purposes, building law and regulations, administration and control of building contracts, carrying out structural surveys and advising on repairs, energy conservation, and many other matters in both the public and private sectors. This is a growing profession, dealing with a wide range of legal and technical problems.

c) The land agent or agricultural surveyor

traditionally works in rural surroundings. The work includes the valuation, sale, and management of rural property and the sale, by auction, of live and dead stock. Such a surveyor may also be concerned with forestry, farm management, and rural planning. Besides private practice, employment may be found with bodies such as the National Trust and the Nature Conservancy Council, as well as some rural local authorities.

d) The land surveyor

will map and plan for many varied purposes. He will undertake precise measurement and positioning of construction works, besides monitoring structural movement. He will survey anything from a simple building site, through a proposed motorway route, to a mountainous region. Employment is in private practice, commercial survey firms, and local and national government agencies. He can expect to find work world-wide.

e) The hydrographic surveyor

will undertake similar tasks to the land surveyor but offshore, making charts of the sea bed and surface features within ports, coastal waters, shipping lanes, and more distant deep-sea areas for the purposes of navigation, dredging, offshore gas- and oil-related construction projects, and mineral-resources development. Employment can be with port authorities; naval departments; and dredging, hydrographic, geophysical, oil, and construction companies. Again, employment can be world-wide.

f) The minerals surveyor

will plan the development of future mineral workings. He will advise on planning applications and appeals, mining law and working rights, mining subsidence and damage, environmental effects of mines, and the rehabilitation of derelict land. He manages and values mineral estates and surveys workings on the surface or in deep underground mines. His employment may be in private practice or with multinational concerns or government departments.

g) The planning and development surveyor

will specialise in all aspects of urban and rural planning, working as part of a team and offering advice on economics and amenities, conservation, and urban renewal schemes. Development surveyors work closely with the planners, to implement their plans within a given time scale and budget. Employment is found in both private practice and local government.

h) The surveying technician

will normally join a surveying office straight from school and work under the supervision of a qualified surveyor. Technicians are found in all divisions of surveying and, over the years, can gain excellent practical experience and technical knowledge, becoming an invaluable assistant and member of the office team.

However, as well as those occupations which contain the title of surveyor, there are other professions and callings which require a sound knowledge of the basic principles of surveying. These include architecture, engineering, construction, and so on.

This volume is therefore aimed at those students who will, hopefully, at the end of their studies enjoy a career in the professions, the construction industry, civil engineering, or any allied field. With this in mind, the chapters which follow will be concerned with the basic principles involved in the surveying of land and buildings.

The origins of land survey

The need to survey land first arose when land suitable for cultivation required to be measured out to enable individual plots to be registered.

Early land measurement can be traced in Babylonia and Egypt in the form of field measurements, while the Egyptians also made explorations of the Nile up to its junction with the Bahr. The Great Pyramids, which are still something of a mystery, could not have been built without a knowledge of surveying principles, in spite of the conflicting views as to how these artefacts were constructed physically. One thing is certain, however, and that is that the Egyptians knew how to range out a straight line using wooden rods in exactly the same way as we do today. They also knew how to set out right angles and other angles.

Masterpieces of Greek architecture – the Parthenon, the Erechtheion, and the theatre of Dionysus for example – also demonstrate surveying skills; while the Romans – considered by many authorities to be the greatest of the world's engineers – not only understood and used the principles of surveying but developed them further. Ptolemy's *Geografia* consisted of manuscripts which needed a knowledge of cartography as well as of land-survey methods in order for them to have been produced. The Roman aqueducts at Tarragonna and Nîmes are not hit-and-miss affairs but precise pieces of engineering, requiring great skill and knowledge of surveying for their erection. It should not be forgotten that the greatest tribute to these early skills is the fact that these structures are still with us today, some 3000 years later, and that after this long period of use (and misuse) they are in many cases still capable of doing the job for which they were designed; for example, the *Cloacca Maxima* – the great sewer of Rome – is still in use in various parts of that city.

However, during the so-called Dark Ages, much knowledge was stored away rather than used, and in some cases lost, until rediscovered later, and the development of surveying was no exception. It is safe to say, therefore, that it is the ancient world and the modern world together which have resulted in the complex profession that we know today.

Britain also had its great buildings and its great architects/surveyors who used their knowledge of surveying to the full. Just as the ancient world tied in surveying to astronomy, the same could be said of Sir Christopher Wren – scholar, mathematician, astronomer, architect, and scientist – who for a time held the office of Astronomer Royal. Others such as Leonard and Thomas Digges (father and son), Jonathan Sissons, and Jesse Ramsden all helped to develop the theodolite, while Aaron Rathbone wrote in praise of the plane-table method of surveying, besides developing the surveyor's chain.

It is from the principles of the Egyptians, Greeks, and Romans that men such as Galileo, Leonardo da Vinci, Marco Polo, Columbus, Drake, Digges, Rams-den, and others were able to pave the way for the development of the techniques and instruments now available to the modern surveyor.

It is also apparent from our brief trip through history, that land survey has many parts, which can be listed as

a) the measurement of land,
b) the setting-out of buildings,
c) the development of routes and the production of maps,
d) the measurement of buildings and other artefacts,
e) the calculation of areas and volumes, and
f) the measurement of angles.

and it is to these areas that the student's studies will be directed.

Land surveying as it affects the construction industry

This aspect of surveying can be considered in three ways:

i) the straightforward tasks easily carried out by members of the construc tion industry,
ii) those tasks carried out by members of professions allied to the construc-tion industry,
iii) the more complicated and intricate tasks requiring the services of a land surveyor.

Although the skills of architects and surveyors during the past century or so have played a considerable part in the great changes that have taken place in almost every corner of the globe in land and property planning, use, and development, the greatest changes within the field of land survey have occurred during the last few years. There is no doubt that this trend will

The Pont du Gard at Nîmes (completed about 19 B.C.) rises 49 m above the valley floor. The arched construction was determined by the position of rocks in the river bed.

St Paul's cathedral in London stands 111 m high and was built in 35 years, from 1675 to 1710.

continue, and there is also no doubt that the major breakthrough has revolved around one word – electronics.

Electronic equipment can accomplish tasks in half a day that would take up to three weeks by conventional methods. It can perform tasks which would hitherto have been impossible or at best have yielded unreliable and unsatisfactory results.

However, the student must not, in view of these developments, consider that learning the simple basic and time-honoured methods is a waste of time, for it is from these basics that the more sophisticated techniques have evolved. Other reasons must also be borne in mind:

a) Most land survey tasks are of a simple nature, and sophisticated and expensive equipment will be neither necessary nor available. Equipment should be suited to the nature of the task and the expertise of the operative.

b) The small architect's or surveyor's office, carrying out everyday jobs, will not have the staff or the equipment available to carry out work on a grand scale, which would be the domain of the specialist surveyor.

c) Clients do not have the financial resources to pay for the unnecessary. The development of four houses on a small plot of land does not call for expensive electronic equipment. By the same token, the survey of a mountainous region would not be done by chain and tape. In other words, the equipment used must be selected to suit the task in prospect.

d) Generally, contractors can handle most tasks on site (from setting-out to plumbing steelwork) by use of simple equipment and operatives trained in the basic principles.

It is, therefore, the intention to keep things as simple as possible in this book so that, armed with a knowledge of the basic principles, the student should be of use in the professional office or on the building site at the end of a year's course of study.

Surveying, as a subject, is not difficult and, like any other subject, common sense and a feeling for the topic are the major requirements. Although a good basic mathematical background is an advantage, the subject can be learned provided that attention is paid to each step of the way. Short cuts do not form any part of surveying and must be resisted if success is to be achieved.

1

Basic Principles

Before becoming involved with the instruments used for survey work, it is necessary for the student to understand *why* surveying is carried out, *what* type of survey can be undertaken, and *when* one technique will be more suitable than another for the task in hand.

The terms *surveying* and *levelling* must also be defined if any proper sense is to be made of the subject under consideration.

1.1 Purposes of surveying

The various techniques of surveying are used for two distinct purposes: (i) the measurement of existing land, buildings, and other man-made features and (ii) the setting-out of works. These may be more formally defined as follows.

i) *Land surveying* is a branch of applied mathematics dealing with the measuring and recording of the size, shape, and contents of any portion of the earth's surface and the drawing out of this information, to a suitable scale, in the form of a two-dimensional map, plan, and/or section.

Since land and artefacts are three-dimensional, land surveying also includes *levelling*, which is the determination of the relative heights (altitudes) of the different points in the area under survey. It is this information which, when coupled with that of the plan, enables a section through the land and/or artefacts to be drawn (see Fig. 1.1).

ii) *Setting-out* is the term used for the operations necessary for the correct positioning of proposed works on the ground and their dimensional control during the construction process.

In effect, setting-out is the converse of surveying in that a drawing is produced, and then the measuring is carried out in order to translate the information given on the drawing into fact on the ground.

Once the surveying has been completed, the resulting drawings will form the basis upon which the designer – be he architect or engineer – will prepare his design and working drawings of the proposed works. It is most important that the survey is done accurately and carefully, so that any proposals will fit on to the land and into the position required.

A third element relating to both land surveying and setting-out is the calculation of areas and volumes. This aspect is especially relevant where areas of land require to be cut out or filled in and the volumes of earth or other material needing to be carried away or transported to the site must be known.

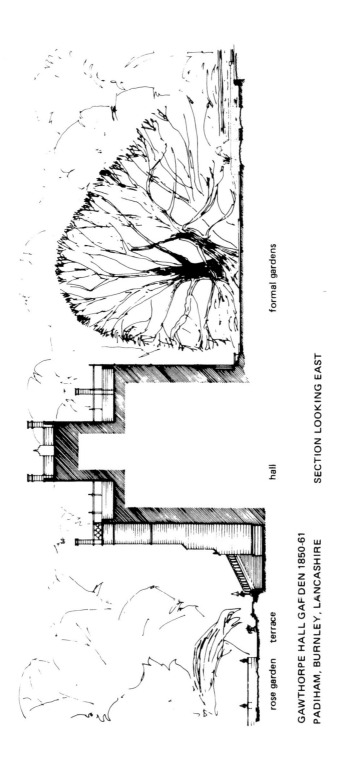

GAWTHORPE HALL GAFDEN 1850-61
PADIHAM, BURNLEY, LANCASHIRE

SECTION LOOKING EAST

rose garden   terrace   hall   formal gardens

Fig. 1.1 Section through land and building

In summary, the term *surveying* is often confined to land measurement in the *horizontal plane* (and to buildings generally) while the term *levelling* is used to indicate those operations which fix heights or differences in level (altitude).

1.2 Branches of surveying

The earth is an *oblate spheroid*; that is to say it is a sphere flattened at its poles and having the shape of an orange. However, for surveying purposes, the earth can be regarded as a true sphere of some 12 751 km *mean diameter*, which gives rise to the following definitions:

a)   A *level surface* (or level plane) is a surface or plane partaking of the curvature of the earth. It is *not*, therefore, the *flat* surface or plane met with in the study of plane geometry but is *curved*.

b)   A *level line* is any line lying in a level surface.

c)   A *horizontal plane*, at any point, is *tangential to the level surface at that point*.

d)   A *horizontal line* is any line lying in a horizontal plane.

In Fig. 1.2, a level surface is illustrated by the shaded portion of the surface of the sphere, with AB a level line within such a surface. The horizontal line is shown tangential to point A of the level surface. (Further discussion is to be found in Sections 4.3 and 5.1.)

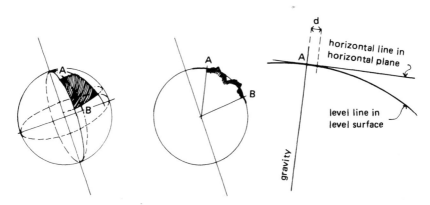

Fig. 1.2 A *level surface* is not the flat surface or plane met with in plane geometry – it is curved.

It is fairly obvious from the above that the *curved* surface of the earth cannot be accurately represented on a *flat* sheet of drawing paper, which can only effectively represent the horizontal plane. How, then, can a survey map or plan be a two-dimensional representation, to a suitable scale, of the earth's surface? Because of the very large mean diameter of the earth, its curvature can be ignored up to certain arbitrary limits, and for surveys of limited extent, there is no appreciable difference between measurements assumed to be on a plane surface and those made on an assumed spherical surface. Such an arbitrary limit is shown in Fig. 1.2, where, for the distance *d* (from point A), the horizontal plane and the level surface are considered to be coincident.

This fact has given rise to two principal classifications of surveying which are

    i) *Geodetic surveying*, where measurements cover such a large part of the surface of the earth that the curvature of the earth *cannot* be ignored if discrepancies are to be eliminated. (Note: On the surface of a sphere the angles of a triangle do not add up to 180° as in plane geometry.)

    ii) *Plane surveying*, where, because of the limited extent of the survey, the curvature of the earth *can* be ignored. Depending upon the accuracy required, the area which can be considered as a flat plane is up to 300 km².

Within these classifications, the principal types and purposes of surveying are

a) *Geodetic surveys* — high-accuracy surveys which are concerned with the shape of the earth or the position-fixing of control points for use in the carrying out of surveys of a lower accuracy.

b) *Topographical surveys* – the location of the main natural and artificial features of the earth (including hills, valleys, lakes, rivers, towns, villages, roads, railways, etc.).

c) *Cadastral surveys* – surveys for legal purposes such as deed plans showing and defining legal property boundaries and the calculation of the area(s) involved.

d) *Engineering surveys (including building surveys)* – preparatory to, or in conjunction with, the carrying out of engineering or building works (roads, railways, tunnels, dams, reservoirs, sewerage works, drainage, buildings, etc.). Such surveys would include setting-out drawings.

e) *Military surveys* – usually carried out for defence purposes. It is interesting to note that the United Kingdom Ordnance Survey originated from the need for a military survey of the country. The organisation is now a civilian one, responsible for producing up-to-date survey maps of the UK (except Northern Ireland) (see Section 1.5.8).

There are, of course, other purposes such as reconnaissance surveys, location surveys, geological surveys, exploratory surveys, and many others of a more minor nature too numerous to mention. All, with the exception of the geodetic survey, can however be accomplished using simple plane-survey methods and, since this volume is concerned with surveys whose size will generally be within the previously mentioned arbitrary limit of 300 km², no further consideration will be given to the rather advanced work of the geodetic survey.

1.3 Branches of technique

Having discussed the objects and purposes for which a survey is carried out, consideration must now be given to the techniques by which the necessary information can be acquired in order that a drawing may be produced. The principal branches of technique are

a)   chain surveying,

b)   plane-table surveying,

c)   traverse surveying,
d)   tacheometric surveying,
e)   triangulation surveying,
f)   hydrographic surveying,
g)   aerial surveying.

Techniques (a) and (b) will be covered in detail, with (c), (d) and (g) being touched upon briefly. The remainder are outside the scope of this text, but the student should nevertheless be aware of their existence.

1.4 Basic survey methods

In order fully to understand the above techniques, an appreciation of the methods of obtaining the required information is necessary. By ignoring instruments and equipment, the techniques can be broken down into *four basic methods* of collecting information. Each method relies on the simple principle that if *two* points are established, a *third* point can be located in relation to them by various forms of measurement.

a)   **Linear measurement** – measurement having only one dimension, i.e. length. Such a measurement in a straight line would give the shortest distance between any two points. When two linear measurements are multiplied together, square measure or area results (see Chapters 2 and 3).
b)   **Angular measurement** – the measurement of the angle formed when two straight lines (or directions) meet (see Chapter 6). Although an angle possesses magnitude (i.e. size), it cannot be estimated as a length, breadth, or area; therefore special units are used, i.e. degrees and radians.

In order to discuss the methods in detail, it is necessary to state the following:

i)   *The situation*   In Fig. 1.3, the line AB represents a straight wall, while C is a point (say a vertical metal post) some distance way.
ii)   *The requirement*   To produce a two-dimensional plan, drawn to some suitable scale, showing the post in true relationship to the wall. (Note: Drawing out the measured information is known as *plotting the survey*.)
iii)   *The problem*   How can the post be located by measurement in relation to the wall in order that the requirement may be fulfilled?

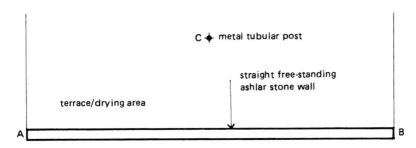

Fig. 1.3 The situation: metal post in relation to straight stone wall

(Note: It is a prerequisite of each of the following methods that a freehand sketch of the situation be made, not to scale but of reasonable proportions, upon which the measured information and any relevant notes will be written in a clear readable manner. It is also to be understood that all linear measurements will be made in the *horizontal plane*, thus enabling a *true plan* (i.e. a true projection) of the ground to be drawn. (See also Section 2.4.6))

1.4.1 Method 1 – intersecting arcs

i) *On site* Measure the horizontal distances AB, AC, and BC and note down the information on the sketch (Fig. 1.4(a)).
ii) *In the office* Draw line AB to scale. Using compasses, swing an arc from A with the radius set to the scale length of AC. Similarly, swing an arc from B with the radius set to the scale length of BC. The intersection of these arcs will locate point C (Fig. 1.4(b)).

This method is the basis of *chain-survey technique* and may be used for land survey, building plans, etc. as will be discussed in Chapters 2, 3, and 7.

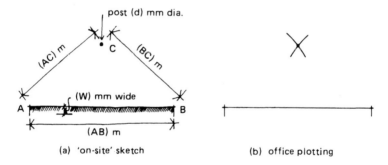

(a) 'on-site' sketch   (b) office plotting

Fig. 1.4 Intersecting arcs

1.4.2 Method 2 – offsets/co-ordinates

i) *On site* Measure the horizontal distance AB. Measure the perpendicular distance from point C to the wall (this will create point D) and then measure the distance from one end to the point D (either AD or BD – surveyor's decision). Note down the information on the sketch (Fig. 1.5(a)).
ii) *In the office* Draw line AB to scale and scale off the distance AD (or BD) from the end of the wall. Set up a right angle at point D by means of compasses or a set square. Scale off the distance CD to locate point C (Fig. 1.5(b)).

This method is known as *offsetting*, and the distance CD is called an *offset*. Note that the measurement is taken *from point C* to the wall, since the required right angle is more accurately obtained this way, as will be shown in Chapter 3. The method is not used to locate important points but to pick up points of detail from main survey lines or detail lines (see Fig. 2.1) in both chain survey and other techniques.

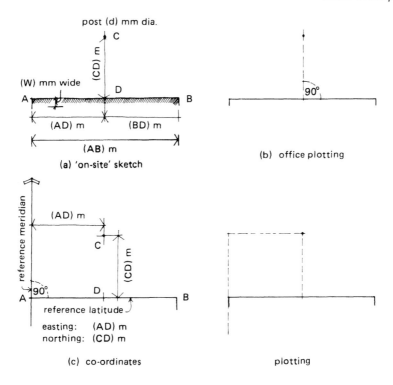

Fig. 1.5 Offset/co-ordinates

In more advanced work, the measurements AD and CD would be known as co-ordinates, with AD being an *easting* and CD a *northing*. A *reference meridian* would pass through point A at right angles to AB, while AB itself could act as the horizontal axis known as the *reference latitude* (Fig. 1.5(c)). This is analogous to locating a point on a graph.

1.4.3 Method 3 – polar co-ordinates/radiation

i) *On site* Measure the horizontal angle BAC the length AB and the horizontal distance from point A to point C. Note down this information on the sketch (Fig. 1.6(a)).

ii) *In the office* Draw line AB to scale. Use a protractor or an adjustable set square to set off the angle BAC. From A, scale off the distance measured to locate point C.

This method is the basis of *traverse survey* – a technique of measuring the lengths of connected lines and the angles between successive lines. There are two types of traverse survey, as illustrated in Fig. 1.7:

a) the *open traverse* (also known as the *unclosed traverse*), where the starting and finishing points are unrelated. This type of survey would be used for long narrow situations, e.g. routes for pipelines, railways, etc., and for reconaissance work. It is not self-checking and the only real check which can be applied is to resurvey the traverse in the opposite direction from E to A.

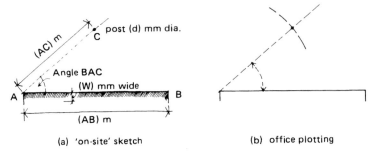

(a) 'on-site' sketch   (b) office plotting

Fig. 1.6 Polar co-ordinates

b) the *closed traverse*, where the traverse starts and finishes at the same point or where the start and finish points are known, being control points of a larger survey.

The student may come across other names such as *polygonally closed traverse* and reference can be made to the function of the traverse, e.g. *control traverse*, *motorway traverse*, or even the type of instrument used to carry out the work, e.g. *compass and chain traverse, theodolite traverse, E.D.M. traverse, etc.*

A traverse survey, often termed simply a *traverse*, may be done quickly using a *prismatic compass* (or *box sextant*) and *chain*, or, if a high degree of accuracy is required, a *theodolite* would be used. In all cases, a *reference direction* is required, from which the bearing of the first line is taken. This, and the use of the compass and theodolite for measuring angles, are discussed in Chapter 6.

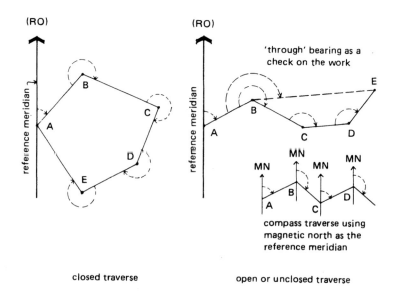

closed traverse   open or unclosed traverse

Fig. 1.7 Types of traverse survey

1.4.4 Method 4 – intersection/triangulation

i) *On site* Measure the length of AB and the horizontal angles BAC and ABC. Note down the information on the sketch (Fig. 1.8(a)).
ii) *In the office* Draw line AB to scale. By means of a protractor or an adjustable set square, set off lines from A and B at the measured angles BAC and ABC. The intersection of these lines will locate point C.

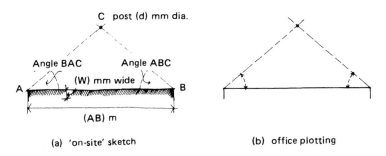

(a) 'on-site' sketch   (b) office plotting

Fig. 1.8 Intersection/triangulation

This method is termed *triangulation* and is the one most frequently used in *precise geodetic surveys* where linear measurement is reduced to the minimum. In precise surveys, however, the plotting would not be done by means of a protractor or an adjustable set square: the triangle would be resolved by trigonometry, using the sine rule, and would be plotted by means of co-ordinates.

It is worth noting at this point that, until the introduction of electro-magnetic distance-measuring instruments which made possible the measurement of the sides of large triangles, it was easier to measure angles more accurately than distances. Precise survey work therefore previously relied on knowing one side length of a triangle and three accurately measured angles.

Where, however, all the sides are measured in high-order work, the technique is called *trilateration*. Neither of the terms triangulation and trilateration as used in high-order work should be confused with the simple linear measurement employed by the technique of chain survey.

While the above methods of taking measurements in the horizontal plane form the basis of all survey work necessary for the production of a map or plan, three of these methods also form the basis for levelling operations.

Ordinary levelling is carried out on the basis of method 2 (Fig. 1.9). AB is now considered to be a horizontal surface and point C a higher altitude point. Once a horizontal line has been established, the vertical distance from point C to the plane AB can be measured.

Trigonometrical levelling makes use of methods 3 and 4 (Fig. 1.10).

All the methods discussed use the most elementary geometry and trigonometry, and the student should have no fear of this provided that the principles are learned thoroughly and the temptation to take short cuts or become

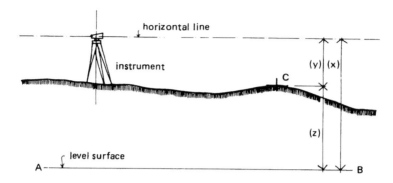

Fig. 1.9 Ordinary levelling. The difference in height (x) between the level surface and the horizontal line can be calculated and the vertical distance (y) from the horizontal line to point C can be measured. The height (z) of point C above AB is given by $(x - y) = z$.

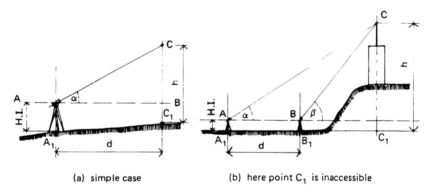

(a) simple case   (b) here point C_1 is inaccessible

Fig. 1.10 Trigonometrical levelling. The *theodolite* (see Chapter 6) may be used to determine the relative heights of two points. In (a), only one vertical angle is observed, because both points are accessible. In (b), two observations are necessary since only one point, C, is visible.

slipshod in approach and practice is resisted. Armed with these principles, the student should find little difficulty in translating full-sized fact into a scale drawing and, with experience, will be able to adapt the methods to suit his needs with every confidence of a successful result.

1.5 Conducting a survey

There are normally two ways of doing most things – the right way and the wrong way – and surveying is no exception to this rule. It is also a fact of life that once things are learned incorrectly it is very difficult to 'unlearn' them and relearn them correctly. The following guide-lines will, if borne in mind at all times, ensure success in field-survey work. They are so important that they will be further expanded throughout the book.

1.5.1 Working from the whole to the part

In a chain survey, for example, the main frame must be set out first and the specific detail then be added on to the frame – not the reverse. There is far less error when a large area is divided into smaller units than when small units are surveyed and an attempt is made to join them together to form the large area. It is very much like doing a jigsaw puzzle: most people have more success by sorting out all the straight bits and making the frame first. Within this overall confine they can relate specific areas to the picture on the box, knowing the size of the finished puzzle. Similarly with a survey: once the overall size has been determined, the smaller areas can be surveyed in the knowledge that they must (and will, if care is taken) fit into the confines of the overall frame, just as the pieces of the jigsaw puzzle fit into the frame without any pieces being spare.

1.5.2 Checks on measurements

Checks should be arranged on all measurements wherever possible. It is always preferable to take a few more dimensions on site to ensure that the survey will resolve itself at the plotting stage, rather than to have to return to the site to take more measurements because things do not tie in on the drawing-board. Returning to site can often be quite an expensive business, besides the frustration and loss of time it can cause.

1.5.3 Field notes

These must be *clear* and *complete*. Never trust to memory – it can play tricks. The survey may have to be drawn out by a draughtsman who was not involved in the physical survey operations, therefore *all information must be on the survey notes* and not in someone else's head.

1.5.4 Honesty

Honesty is essential in booking notes in the field and in plotting and calculation in the office. There is absolutely nothing to be gained from 'cooking' the survey or altering dimensions so that points will tie in on the drawing. Others will undoubtedly be basing their own work upon your results and putting their trust in those results and the surveyor who produced them. It would be utterly unprofessional to betray that trust.

1.5.5 Concentration and care

In order to ensure that all necessary measurements are taken to the required standard of accuracy and that nothing is omitted, concentration and care must be maintained in the field at all times. This applies to the assistant equally as it does to the surveyor in charge. For instance, even on the best organised surveys, there may be a lull in proceedings while the surveyor sorts out the unexpected – this is not an opportunity for the assistants to practise javelin throwing with the ranging rods. Assistants must also listen carefully to all instructions and carry them out to the letter *without question*. There can be but one boss on a survey, who will take full responsibility for all things, and, if success is to be achieved, concentration and care must run from top to bottom of the team.

1.5.6 Equipment

The equipment selected should be appropriate to the task in hand. A steel chain, giving at best an accuracy of 1/500 to 1/1000, would be of little use for work requiring an accuracy of 1/10 000. Similarly, a theodolite reading to one second of arc would be pointless where a reading to one minute is sufficient.

Having selected the equipment necessary to the work, it should be thoroughly checked and, if found wanting, should either be adjusted, repaired, or replaced or have allowances calculated for its deficiencies. This task will be less tedious if field equipment is regularly maintained. This maintenance should be augmented by proper protection in the field and by cleaning after use. *Always leave equipment, after use, in the state that you would like to find it* – ready for immediate use after an initial check.

1.5.7 Drawing

Draughtsmanship should be of the highest order, both in the field notes and in the final office plotting, and information including the scale and the units used, a north point, map references, the date, the surveyor's name, the draughtsman's name, the drawing number etc. is vital for future reference.

Any two-dimensional map, plan, or building drawing to a recognised scale is a proportional representation, on paper, of actual features on the ground or proposals which may become fact. The *ratio* of a dimension on a drawing to the equivalent full-size measurement on site is known as the *scale* of the map or plan.

For example, referring back to the situation shown in Fig. 1.3, the length of the wall AB on the finished drawing is 100 mm (0.1 m). The actual measurement of the wall on site is 50 m. The ratio of the two dimensions to one another is therefore 100/50 000 (0.1/50) or 1/500. The drawing is therefore drawn to a scale of 1/500.

In effect this means that 1 mm on the paper represents 500 mm on the ground. The scale would be noted on the drawing by the convention 1/500 or 1:500, which is known as the *representative fraction* of the drawing.

The choice of scale for the finished drawing will be determined at the pre-survey stage and will depend on the purpose for which the survey is being made. (This purpose will also have a bearing on the standard of accuracy required in the field.) Basic scales can range from 1:1 to 1:1 000 000, the larger scales being used for detail drawings and plans while the smaller ones are used for mapping, as can be seen in the following guide (where O.S. = Ordnance Survey):

a) architectural work, component and assembly detail drawings, working drawings, and location drawings – 1:1, 1:5, 1:10, 1:20, 1:50, 1:100, 1:200

b) civil-engineering works, site plans and key plans, surveys, and layouts –
    1:500, 1:1000, 1:1250 (O.S.), 1:2000, 1:2500 (O.S.)
c) town surveys, highway and route surveys – 1:2500 (O.S.), 1:5000, 1:10 000,
    1:20 000, 1:50 000
d) maps – 1:25 000, 1:50 000, 1:100 000, 1:200 000, 1:500 000, 1:1 000 000

Although scales may be *large* or *small* there is no definite dividing point. In other words, one scale may be said to be larger than another. The simple way for the student to work this out is to remember that the bigger the denominator of the representative fraction, the smaller the scale of the drawing. This can be seen as follows:

1:1, where 1 mm on the drawing represents 1 mm on the ground, is a
    larger scale than
1:50, where 1 mm on the drawing represents 50 mm on the ground, which
    in turn is a larger scale than
1:200, where 1 mm on the drawing represents 200 mm on the ground, and
    so on.

The distinction between maps and plans is easily understood. A *plan* is a true-to-scale representation of what is either on the ground or what is proposed. A *map* is drawn to such a small scale that some of the features on it cannot be drawn to scale. Rivers and roads may be shown by the thinnest possible line but will be disproportionate if they are to be shown at all on small-scale maps. For example, on a 1:1 000 000 map, a road 5 m wide *should* be represented by a line of 0.005 mm thickness. This would not show up, since it is too fine to be drawn; therefore, if the road is to be shown, a thicker line will have to be used which obviously will not be to scale.

1.5.8 Ordnance Survey

The Ordnance Survey (O.S.) was founded in 1791 from already mentioned military origins. As the official UK map-maker, O.S. now produces an increasing range of maps, atlases and guidebooks providing information for architects, surveyors, town planners, essential services such as Fire, Police and Ambulance, motorists, holiday makers, cyclists, climbers and ramblers – in fact for any person or organisation concerned with the design and enjoyment of the town and country landscapes of Britain. The study of O.S. maps is a complex matter and the student is advised to make reference to the handbook and general information published by the O.S. itself.

This text, however, is concerned mainly with O.S. maps of the scales

1:25 000; 1:10 000; 1:2500 and 1:1250

which form part of the *national grid series* of maps. The principal matters which make a map useful for its purpose are

i) *its scale* which, being a ratio, will allow metric or imperial measurements
    to be read by using the appropriately divided scale-rule;
ii) *the detail* which it includes, with the larger scale maps having the greater
    detail.

a) *the use of maps* will vary with the scale and the following text gives a brief account of each of the four maps mentioned above, along with an indication of some possible uses.

   i) *1:25 000* is a relatively small scale map and the smallest scale at which field boundaries are shown. Contour lines are drawn at 10m V.I. (see Section 5.12.1(c)) with altitudes shown as spot-levels at regular intervals along main roads.

Maps at this scale would be used

- when planning large-scale engineering works involving gradients of roads, sewers and pipes;
- extensively when involved with the flow of rivers and streams (flooding abatement, irrigation, reservoirs);
- where the contour lines provide a means of solving problems of intervisibility and clearance between points;
- to illustrate aspects of regional planning because the small scale enables an 'overview' to be given.

   ii) *1:10 000* is almost accurately drawn to scale although some road widths are increased to accommodate road names. Conventions (signs and symbols) are used to represent features in a semi-pictorial manner, e.g. orchard, quarry, cutting, embankment, etc. (see Fig. 3.18), whilst individual parcels of land are shown, together with fences and fields. Contour lines are drawn at 5m V.I., although this is increased to 10m V.I. in mountainous areas.

Maps at this scale would be used

- by the surveyor involved in estate management because individual tenant holdings can readily be distinguished;
- for design of schemes for water supply;
- for geological surveys;
- by town planners and urban designers to illustrate initial proposals.

   iii) *1:2500* is a highly detailed map providing accurate information to a fairly large scale. A distinctive feature of the map is that each parcel of land is identified by a number and has its area printed below (in hectares and acres) which makes the map extremely useful for rating and valuation purposes as well as location plans for Local Authority submissions.

   iv) *1:1250* is the result of a double enlargement of the 1:2500 sheet which renders it no more accurate than the smaller map. It is the largest scale of mapping published by the O.S., although in the 19th century and early 20th century 1:500 scale maps were produced and are still to be found in many offices. At 1:1250 scale all streets are named, as are public and other buildings having a specific name. Remaining buildings are numbered.

Maps at this scale are used

- in part, as block plans or location plans when making applications for Planning and Building Regulation approval;
- by designers for initial layouts;
- by statutory undertakings to record the positions of power lines,

buried cables, sewers, drains, water pipes, gas pipes, telephone conduit etc.;

- by Town Planners and Environmental Officers when dealing with clearance areas, redevelopment areas, etc., because individual buildings/dwellings can be indicated in terms of condition by using different colours.

On maps of this scale, as on the smaller 1:2500, there is much blank space which can be used for the addition of information.

b) *reproduction of O.S. sheets*, in whole or in part, can only be done under licence, although Local Authorities can, at their discretion and with the agreement of the O.S., allow non-professionals to purchase small areas of 1:1250 plans showing their property for use when Planning and/or Building Regulations approval is being sought.

1.6 Accuracy and precision

Accuracy allows a certain amount of tolerance, either plus or minus, in a measurement; while precision demands exact measurement.

Since there is no such thing as an absolutely exact measurement, plane-survey work is usually described as being to a certain *standard of accuracy* which in turn is suited to the work in hand. Bearing in mind the purpose for which the survey is being made, it is better to achieve a *high degree of accuracy* than to aim for *precision* (exactness) which, if it were to be attained, would depend not only on the instruments used but also on the care taken by the operative to ensure that his work was free from mistakes.

Using the standard chain and exercising reasonable care, an accuracy of 1 in 1000 should be attainable in open country. In rather more difficult terrain (hilly or broken ground) an allowable error would be 1 in 500. Using a steel band and taking special precautions, an accuracy of 1 in 5000 should be possible.

Always remember, however, that the greater the degree of accuracy required, the greater the effort and time needed both in the field and in the office, and of course the more expensive the survey will be for the client.

The standard of accuracy attained in the field must be in keeping with the size and purpose of the survey and the scale of the ultimate drawings.

It is accepted that the smallest measurement capable of being plotted by eye on paper is about 0.2 mm. It could therefore be pointless to take field measurements to a degree of accuracy which would be smaller than 0.2 mm when drawn to scale. For example, on a drawing at a scale of 1:500, 0.2 mm on the paper would represent 100 mm on the ground. It would be sensible, therefore, to book a field measurement of 1.525 m as 1.500 m and ignore the unplottable measurement of 0.025 m (25 mm).

Some authorities would suggest limits of measurements in the field related to the scale of the final drawing:

Scale	Measure detail to nearest
1:50	0.005 m (5 mm)
1:100	0.010 m (10 mm)
1:200	0.010 m (10 mm)
1:500	0.100 m (100 mm)

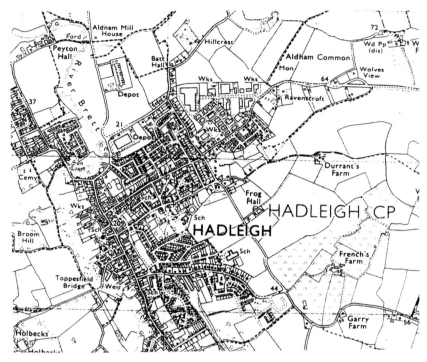

1:25 000 © Crown Copyright

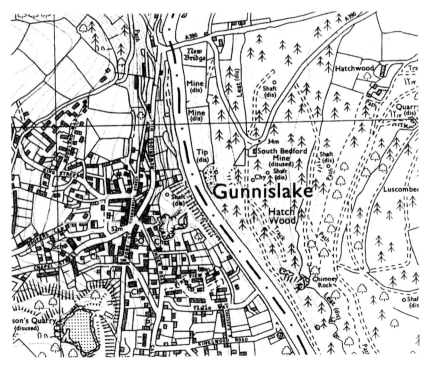

1:2500 © Crown Copyright

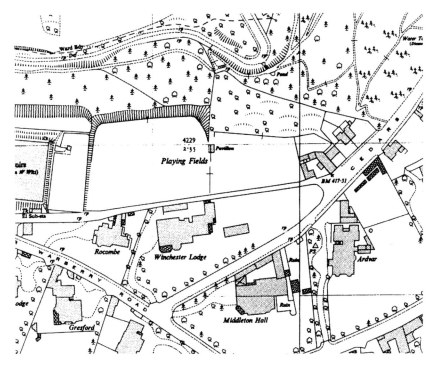

1:10 000 © Crown Copyright

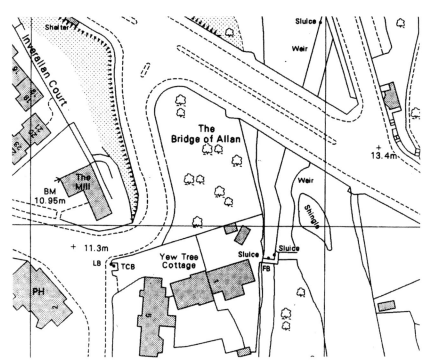

1:1250 © Crown Copyright

There is, however, a danger of becoming slipshod or lazy in the field, and it is better to book the complete figure, e.g. 1.525 m, even though this may ultimately be plotted as 1.500 m. This would act as a safeguard should a larger-scale 'part drawing' be required unexpectedly at some future date.

While it may be argued that it is much faster in the field to read and book 1.500 m rather than 1.525 m, *this method could not be adopted for main survey lines* or objects whose actual dimensions are required for purposes other than plotting. In view of this, it makes good sense to book down the actual dimensions in full and make any adjustments at the plotting stage.

Care must also be taken when taking *Offsets* (see Section 3.2). Longer offsets may be taken in the field for plotting at a small scale and the survey will be within the plotting accuracy of that scale. However, for large scale plotting longer offsets may not be sufficiently accurate and additional precautions such as *ties* may be necessary.

Finally, an accuracy of 1 in 100 000 is not normally expected, or indeed attainable, in plane-survey work; while an accuracy of 1 in 200 is far too low to be acceptable in any kind of work.

1.7 Errors

Although errors and their resolution will be dealt with in detail in later chapters, it will be useful to consider them here in general terms as a form of introduction to them.

No measurement in a survey is ever *exact*, and every measurement – whether linear or angular – will contain some form of error. It is obvious that errors should be kept to an absolute minimum by careful work, but to do this it is necessary to appreciate not only the types of error but their relative importance to the task in hand.

Basically there are three types of error which must either be minimised by extra care, calculation, and correction or be accepted as of little importance.

1.7.1 Mistakes or gross errors

These are serious mistakes by either the surveyor or his assistants – either singly or in combination – and mostly there is no excuse for them. Examples include

a) incorrect booking of a measured length – e.g. writing 6.520 m instead of 16.520 m;
b) incorrect reading of the staff when levelling – e.g. reading 2.340 instead of 3.340;
c) moving the staff position at a change point (see Section 5.1.9);
d) measuring to the wrong point;

and so on.

These can be eliminated only by proper and careful methods of observing and booking and constantly checking both operations. The greatest danger is where readings are taken by one person then called out to a second person for booking. This second person should *always* call out the information again as he has it in his book, for confirmation by the reader, before progressing to the next point.

1.7.2 Systematic or cumulative errors

These are errors which always recur in the same instrument or operation. Their effect is cumulative, which means that this effect increases as the survey progresses.

For example, if a nominal 20 m chain has been stretched, through use, by 50 mm, every time the chain is laid down there will be an error in the distance measurement of 50 mm (0.05 m). This means that, when the chain has been laid down five times, the length will be noted in the survey book as 100 m but the true distance measured will be 100.250 m. The error will have accumulated to 5 × 50 mm = 250 mm.

This type of error would be regarded as *negative*, since its effect is to make the measured length appear *shorter* than the true length. Conversely, if the chain were less than 20 m, due to bent links etc., then the cumulative error would be regarded as *positive* since its effect then would be to measure the line as being *longer* than the true length.

These errors can be minimised by using suitable operational methods and by applying appropriate corrections to the actual measurements.

1.7.3 Accidental errors

These have nothing to do with 'accidents' in the field, such as an assistant falling down a hillside or from a ladder. Accidental errors are those errors which are caused by variations in the eyesight of persons using instrument telescopes, sudden changes of temperature or wind, slight imperfections in the instruments, atmospheric disturbances, and the like. They are usually small, can be either negative *or* positive, are not usually cumulative in the normal sense, and, since they very often cancel themselves out (are said to be 'compensatory'), they are of little importance and can be ignored.

The surveyor's efforts, then, should always be directed at eliminating mistakes, keeping systematic errors to a minimum, and being aware of the possibility of accidental errors, while striving to achieve the standard of accuracy required by the task in hand.

1.8 Control

In surveying terms, the word 'control' can have several meanings which will range from the control the surveyor has over his personnel in the field to the control met with in the setting-out of works (see Chapter 8).

Control can be exercised by the survey itself, in the form of control points to which other aspects of the work must relate. The use of maps is often necessary when selecting control points, which may be Ordnance Survey triangulation points (see Section 5.12.2(j)), bench-marks (Section 5.1.3), or other features.

The whole aspect of control is embodied within the basic principles of surveying and will be referred to throughout the text.

1.9 Summary

Having considered in some depth the underlying principles of surveying, the student should have sufficient understanding to be able to undertake the proper study and practice of an individual survey technique without much difficulty. However, before doing so, the following sequence of operations should be thoroughly mastered, since it will apply whatever the task.

i) Determine the reasons for which the survey is required and the purpose to which the finished drawings will be put on completion. (The survey may be for record purposes and the final drawings for exhibition; or the survey may be for development purposes with drawings forming the basis for further drawn work, etc.)

ii) The site (or buildings) must be inspected and possibly a reconnaissance survey be carried out in order to foresee any inherent problems.

iii) Once the answers to the questions embodied in items (i) and (ii) have been formulated, the most suitable technique and the standard of accuracy required can be selected.

iv) The scale and size of the final drawings will now be determined.

v) The equipment and the instruments selected for the task must be checked for accuracy and working order. Once this has been done, the fieldwork can be carried out.

vi) The last stage will be the production of the final drawings. These will be plotted in exactly the same sequence as the operations were carried out in the field, being drawn out in whole, then in part, as described during the above discussion on basic principles.

Exercises on chapter 1

1. *State* the two distinct purposes for which land is surveyed.
2. *Define* the following terms: (a) land survey, (b) levelling, (c) setting-out.
3. a) What is the difference between geodetic surveying and plane surveying?
   b) Explain the purpose of the following types of survey: (i) topographical, (ii) cadastral.
4. *List* the *seven* principal branches of surveying technique.
5. Explain by means of sketches and notes what are meant by the following survey *methods*: (a) intersecting arcs, (b) offset/co-ordinates.
6. Explain the following terms: (a) trilateration, (b) triangulation.
7. What is meant by the expression 'working from the whole to the part'?
8. a) What is meant by the word 'scale' when related to a map or plan?
   b) A line AB on a survey drawing represents a measured length of 5.000 m. If the actual length of the drawn line is 25 mm, what is the scale of the drawing?

  c) If the 5.000 m length has been represented by a drawn line 100 mm
long, what would then have been the scale of the drawing?
9. What is the difference between (a) a plan and a map? (b) accuracy and
precision? (c) gross errors and systematic errors?
10. *List* the sequence of operations which should be applied to any survey
task.

'How to reduce all sorts of grounds into a square for the better measuring of it' – from
the 1616 English edition of *Maison Rustique, or, the Countrey Farme* by Charles Estienne
and Jean Liebault, translated by Richard Surflet and revised by Gervase Markham.

2

Linear measurement – principles and equipment

The oldest and simplest form of carrying out work within the principal classification *plane surveying* is known as *chain survey*. The principles and methods were known to the Egyptians, and throughout history lines had been measured by means of cords, ropes, or wooden rods, until the seventeenth century when Aaron Rathbone invented the *chain*. This instrument, being the principal item of measuring equipment used, has given its name to the technique.

Chain survey relies directly on the measurement of *horizontal length* (distance) and the *principle of trilateration*. Angles are not measured, except for the occasional right angle or degree of slope, and the equipment ancillary to the chain is very simple and easy to use. Notwithstanding this fact, work of a sufficiently high order of accuracy to cover the requirements of much ordinary engineering and building work is possible.

Quite often the standard of accuracy required is such that, on very small areas, the fibreglass tape or steel band can be used in preference to the chain, but the technique used would still be termed 'chain survey'.

A sound knowledge of chain surveying is therefore essential to a proper knowledge of surveying as a whole, for, in addition to being a complete method in itself, some of the operations of chain survey occur in other branches – notably in traverse surveying.

2.1 Basic terminology

Figure 2.1 shows a typical arrangement sketch upon which the basic terminology has been noted.

2.1.1 Preliminary inspection/reconnaissance

With all surveying tasks, proper preparation before work commences will save time and effort during field operations. The following is a useful sequence.

i) *Consult existing maps/plans* of the area to be surveyed. This will enable a general impression of the area to be gained and may give a clue to any

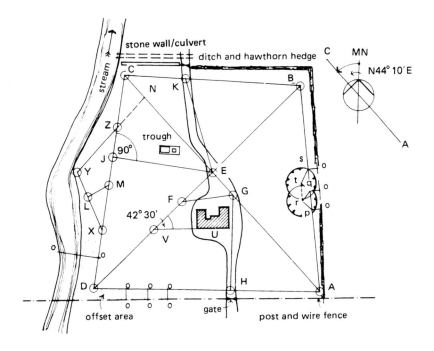

Fig. 2.1 Basic terminology

Chain lines : AB, BC, CD, DA, DB, and AC
Secondary lines : XY, YZ, EK, FG, and GH
Base line : AC supporting main triangles ABC and CDA
Detail lines : KE, FG, GH, and VU
Detail triangle : XYZ, with YZ produced to meet base line at N to provide a check
Perpendicular check line : JE
Diagonal check line : DB
Proof lines : LM, FG, and ZN
Ties : st, tq, qr, and rp
Offsets : oo, op, oq, and os
Offset area : the area between the boundary and the chain lines
Survey stations : A, B, C, D, E, F, G, H, X, Y, Z, and K

    major problems that are likely to be encountered during the surveying process.

ii) *Consider the distance* of the site from the office. What mode of transport will be required or will a separate operational base, e.g. an hotel, need to be organised?

iii) *Locate facilities* on or near the site, e.g. pub, cafe, shops, or will packed lunches and flasks be the order of each day? Remember! you cannot leave equipment unattended in the field while everyone goes off to lunch.

iv) *Contact the owners* of the land/buildings to be surveyed for written permission to enter their property (give dates if possible) and the work required to be done (see Chapter 7).

v) *Walk over the area* to be surveyed (permission to do so should be obtained first) visualising the work to be undertaken, noting any areas which are likely to cause difficulties. These latter can be compared with any problems highlighted at stage (i) above.

A rough sketch of the area should be made in the field-book, showing the survey stations and the routes of the main and other chain lines. It is at this stage that any preformulated ideas (from stage (i)) of a 'modus operandi' are tested in the field and may well be changed. Sight-lines must be checked for obstructions, station points and other points of survey must be accessible and lines joining control points and the detail required must be capable of being measured. Time spent at this stage will save even more time when fieldwork is under way.

vi) *Check all instruments and equipment* thoroughly, not only to see that they are in working order but also to familiarise the survey team in their use.

It is also worth noting at this point that *before any work begins on site*, the owners, tenants, occupiers, etc., should be notified of when work will commence. This may be done prior to the day of commencement or, where people are in occupation, immediately on arrival at the site. Remember, a friendly owner or tenant can be very helpful in providing

(a) safe storage overnight for bulky equipment or for equipment not needed during a day's work,
(b) shelter during sudden showers or storms,

to say nothing of cups of tea/coffee (and even sandwiches) to sustain cold, wet and weary bands of measurers.

2.1.2 Survey stations

A survey station is a point of importance at the beginning or end of a chain line, or at the junction of one line with another. It is usually marked by the insertion into the ground of a vertical ranging pole (see Section 2.3.4). On hard surfaces this point may be marked by a stud, while on normal ground where a more permanent mark is required, a wooden peg (50 mm square) should be driven in, which can be easily located at all times. It is not a bad idea to make a dimensioned sketch of the position of the pegs so that these may be relocated if a peg is lost or accidentally removed. For station points on hard ground which are not to be of a permanent nature, a stand should be used to support the rod vertically.

Stations should be placed as may be found convenient at the corner of areas, or at prominent points, so that the lines joining them are as close as possible to the boundaries of the site in order to keep offset measurements short (see Section 2.1.7).

2.1.3 The base line

This is normally the longest of the chain lines forming the *pattern of triangles*. It should, if possible, be laid off on level ground through the centre of the site and encompass the whole length of the area. A compass bearing should be taken to fix its direction, which in turn will fix the direction of all other lines and allow the position of north to be determined. All survey drawings require a drawn north point.

2.1.4 Trilateration

Chain surveys are built up as a series of adjoining triangles known as the *triangulation pattern*. The surveyor divides the area under survey into a number of triangles which, when all the sides have been measured, can be plotted to scale in plan form. This framework of triangles (see Fig. 2.1) forms the skeleton upon which all the details such as boundaries, buildings, roads, etc. are added after further measurements have been taken.

A triangle is the simplest of all plane figures bounded by straight lines, in that once the lengths of all its sides are known it is completely defined. Any plane figure having four or more straight sides – e.g. a square, rectangle, pentagon, etc. – can be subdivided into triangles by joining opposite corners, and these diagonals can be measured and the triangles be determined without the use of angle-measuring equipment.

When *all the sides* are measured the term used is *trilateration*. This should not be confused with *triangulation*, where *all angles and one side* are measured.

2.1.5 Well-conditioned triangles

All triangles should be 'well-conditioned', i.e. they should not be too acute or too obtuse but roughly equilateral. As much of the area as possible should be included within the large triangles, which in turn should be as few as possible. This has the effect of making the offset area as small as practicable, with offset measurements kept short. It is much better to set up a subsidiary triangle than to measure long offsets.

When dividing the area into triangles, the surveyor will take into account the shape and configuration of the ground and the natural obstacles encountered. Where a well-conditioned triangle is not possible, extra checks should be measured to compensate for this impracticality.

2.1.6 Offset area

The area lying between the chain lines and the boundaries of the site is referred to as the *offset area*. Chain lines are placed as close as is practical to the boundaries so that the area is kept small for calculation purposes (see Section 3.3), as well as to keep offset measurements short.

2.1.7 Offsets

These are measurements taken *from* points on the boundary (or points of detail within the offset area) *to* the chain line. Offsets are kept as short as possible and are measured *at right angles to the chain line*. The exception to this is an *oblique offset*, and care must be taken to indicate it in the field-book along with the angle it makes with the chain line.

Insets are in all respects exactly like offsets but are taken *within the site* and not within the offset area.

2.1.8 Survey lines

These are the main chain lines or other chain lines which are measured.

2.1.9 Check lines or proof lines

When all three sides of a triangle have been measured, although the triangle is completely defined, it still requires to be 'proved'. Figure 2.2 shows examples of how this may be done.

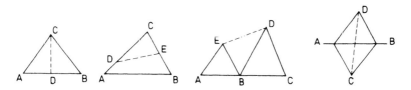

Fig. 2.2 Check or proof lines: lines such as C and D (shown with a broken line) are not used for plotting purposes. The distance is scaled off the drawing and compared with the noted dimension in the field-book.

If one of the lines forming the sides is incorrectly measured, the triangle may still be able to be drawn but it will not represent the triangle as laid down in the field, therefore a check line will prove the accuracy of the fieldwork. Check lines are not used for plotting purposes but are scaled from the drawing and the scaled lengths compared with the measurements taken in the field. In proving a triangle, a check line will 'fix' one angle of the triangle and hence the other two.

In addition to the checks they give, such lines can serve another purpose. They should be arranged so that they can be used to pick up or locate interior detail which cannot be reached from the main chain lines. It is very rare that a check line is a check and nothing more.

2.1.10 Perpendicular check lines

A check line can be laid off from, and perpendicular to, the base line to a survey station point by using the optical square (see Section 2.6.1(a)). This will not only serve as a check line but will also enable rapid calculation of the area of the major triangle involving the base line and station point. The simple formula

$$\text{area of triangle} = \tfrac{1}{2}bh$$

where b = base line and h = perpendicular check line, is easy to use and accurate. This method of calculation should be used wherever possible for determining areas of triangles.

Sufficient check lines should always be measured to ensure accuracy and ease of plotting and calculation, and to pick up all the detail required.

2.1.11 Ties

Ties are either insets or offsets at an angle other than a right angle to a main chain line or other line. They are usually used in pairs to locate some point of detail – e.g. a corner of a building, the centre of a tree, etc. – from two known points on the line. The triangle formed by the ties and a survey line should be well-conditioned for reasons of accuracy (see Fig. 2.2).

2.1.12 Summary

To sum up, the technique of chain survey depends upon a good selection of the framework lines:

a) The base line should be as long as practicable and preferably, although not necessarily, running through the middle of the site area and on fairly level ground.
b) All triangles must be based on the base line – not simply tacked on to each other – or at least be tied back to the base line in some way.
c) All triangles should be provided with checks.
d) Main chain lines should be as close to the boundaries as practicable (but *not* parallel to them), to reduce offset lengths.
e) Lines should be as near horizontal as possible. Over 3° of slope will require allowance calculations to be made (see Section 2.6).
f) Lines should be as free from obstructions and as clear of obstacles as possible.
g) The number of lines should be kept to a minimum, but not at the expense of the accuracy of the work.

Although Fig. 2.1 shows one arrangement for a particular site, no two surveyors think exactly alike and a different arrangement could be used with equal success.

2.2 Equipment and instruments

The equipment necessary to enable a chain survey to be carried out comprises

a) *for linear measurement*: the chain [steel band], arrows, ranging poles, steel tape [fibreglass tape], surveyor's folding boxwood rule [offset staff], plumb-bob [dropping-arrow];
b) *for 90° angle setting-out and slope measurement*: the optical square [prism; cross-staff; box sextant], clinometer [Abney level];
c) *for noting information*: field-book, pencils – HB, F, H – eraser, penknife.

In lists (a) and (b) the items in brackets are those which have equivalent functions to those in the main list and therefore can be used in conjunction with, or as an alternative to, these; for example, the steel band may be used instead of the chain where a higher degree of accuracy is required, while the fibreglass tape may be preferable to the steel variety in wet conditions. The actual equipment taken to the site will largely depend on the surveyor and his appraisal of the work to be carried out.

At least one assistant is required, although it is advisable for the surveyor to have two assistants, so that he can devote his time entirely to directing the operations and recording the information (sketches and noted measurements) in his field-book.

2.3 Equipment used for linear measurement (see Fig. 2.3)

2.3.1 The chain

a) *Rathbone's chain* was used for measuring land and its area. It was 2 or 3 poles in length (1 pole = $16\frac{1}{2}$ feet, about 5 m).

b) *Gunter's chain* was a 4 pole chain – 22 yards or 66 feet long – the length of a cricket pitch. This was a very rapid means of measuring land area, since 10 square Gunter's chains equal 1 acre exactly. There are 10 chains in one furlong and 80 chains in one mile, and these equivalents made the Gunter's chain very convenient for both linear and square imperial measurement. The chain was divided into 100 parts called links, each link measuring 7.92 inches (201.168 mm).

c) The *metric chain* can be obtained in lengths of 20, 25, 30 or 50 metres overall. The 20 m length is the most popular because not only is the reading-and-booking procedure simpler but also this chain is easier to handle in the field than the longer versions. It is divided into a hundred links, each 200 mm in length, the whole unit approximating very closely to the Gunter's chain. Since the standard 200 mm link is common to all metric chains, the 25 m chain has 125 links, the 30 m chain has 150 links, and the 50 m chain has 250 links.

d) The *engineer's chain* – variously called the 'builder's chain' or the 'hundred foot chain' – is of similar pattern to both the Gunter's chain and the metric chain except that each link measures 1 foot. It is heavier to work with and more difficult to align but does not require to be laid down as often as a smaller chain.

Many other chain lengths, now quite obsolete, have been employed at various times (the 'Scots chain', for example, was 22.59 m in length). However, the pattern illustrated in Fig. 2.4, has remained constant for the Gunter's, engineer's, and metric chains.

The chain is robust, easily read, and easily repaired in the field if broken. It consists of a series of dumb-bell-shaped links of steel wire joined together by three small rings for flexibility. The joints are usually left open, but may be welded. At each end of the chain is fitted a brass handle with a swivel joint to the first link. The length of the chain is measured from the outside of the handle at one end to the outside of the handle at the other end, and this length is usually stamped on the brass handles. There is also a swivel joint in the centre of the length to facilitate the folding of the chain, two links at a time, into a bundle somewhat resembling a sheaf of corn and secured by a leather strap (see Fig. 2.3).

To allow fractional parts of the chain to be observed and noted quickly and accurately, brass or plastics tags (or tallies) are attached to the 10th, 20th, 30th, 40th, and 50th link from each end. Since only the five shapes shown in Fig. 2.4

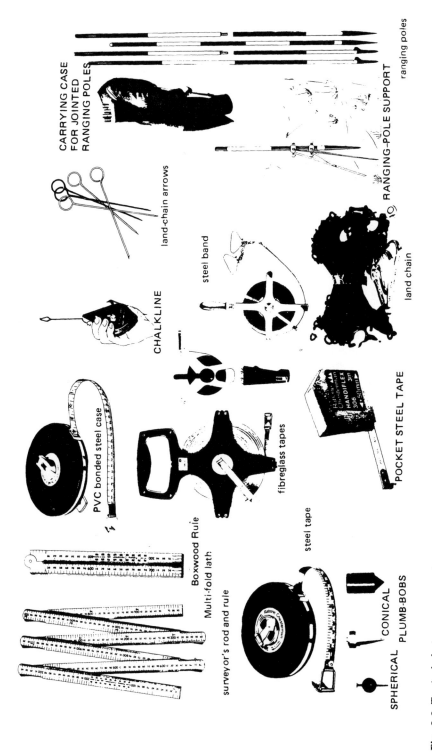

Fig. 2.3 Typical chain-survey equipment

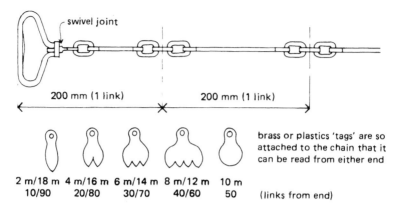

Fig. 2.4 Typical chain pattern. The tags shown are for a 20 m chain and may be of brass or plastics. Alternatively, a plastics tag may be attached at each whole metre position, with a different colour used at each 5 m position. This is usual on chains longer than 20 m and still allows the chain to be read from either end.

are used, the 20 m chain can be read from either end and therefore does not need to be laid down with a specific handle at the starting point.

To lay out the chain on the ground, the strap is removed and the handles are held in the left hand with a few links hanging loose from the main bundle held in the right hand. The bundle is then thrown forwards and the chain, still held by its handles, unfolds itself in a double line. One handle is then handed to an assistant who walks along the double line until the chain is extended to its full length. The surveyor should satisfy himself that there are no kinks or bent links and that the tags are in their proper places.

It is necessary to test the chain from time to time against a 'permanent chain length', and most local authorities will be able to advise on the location of such a facility. If the chain is found to be too short, it is probable that some of the links have become bent and need to be straightened, while over length is more likely due to elongated connections between the links. It is therefore obvious that, in spite of its robustness, undue stress should not be put on the chain during its use, and tugging or violent jerking when moving it or trying to free it from some obstruction will certainly lead to its elongation if not its destruction.

It is not uncommon to find a chain as much as 200 mm too long after abuse in the field, which is why it needs to be checked regularly. However, if the chain is found to be incorrect after measurements have been taken, the following formula will correct the error:

$$\text{true length} = \frac{\text{measured length} \times \text{length of chain used}}{\text{nominal chain length}} \tag{i}$$

Example   A line was measured as 45.800 m with a chain 20.200 m long. What was the true length of the line?

$$\text{True length} = \frac{45.8 \text{ m} \times 20.2 \text{ m}}{20 \text{ m}} = 46.258 \text{ m}$$

If the survey had been measured in links, the formula would still give the correction, as follows.

Example A line was measured with a chain known to be 101 links long and was found to be 229 links. What was the true length?

$$\text{True length} = \frac{229 \times 101}{100} = 231.29 \text{ links}$$

(Note: 231.29 links each 200 mm = 46.258 m.)

If an incorrect chain is used to determine an area, it is not necessary to correct each individual measurement before calculating the area. Compute the area from the measurements taken and then correct this figure by using the formula

$$\text{correct area} = \text{measured area} \times \left(\frac{\text{length of chain used}}{\text{nominal chain length}} \right)^2 \qquad \text{(ii)}$$

This formula derives from equation (i) because an area is given by one length multiplied by another and all lengths are corrected by a constant factor.

Example A plot of ground was measured by a chain 20.200 m long and the area was computed as 1.6 hectares. What was the true area of the plot?

$$\text{Correct area} = 1.6 \text{ ha} \times \left(\frac{20.2 \text{ m}}{20 \text{ m}} \right)^2$$

$$= 1.6 \text{ ha} \times (1.01)^2$$

$$= 1.632 \text{ hectares}$$

Again, if the survey has been measured in links, the formula would still give the correction, as follows.

Example A plot of ground was measured by a chain found to be 101 links long and the area was computed as 1.6 hectares. What was the true area of the plot?

$$\text{Correct area} = 1.6 \text{ ha} \times \left(\frac{101}{100} \right)^2$$

$$= 1.6 \text{ ha} \times (1.01)^2$$

$$= 1.632 \text{ hectares}$$

Finally, after use, the chain should be carefully cleaned and dried, then it should be oiled before being bundled up and stored away. Get into the habit of leaving equipment ready for its next use. Nothing is more frustrating and annoying than to have to set to and clean, or even repair, equipment of any sort before it can be put into service and the survey started – this is time-consuming and costly, at the wrong end of the operation, and a poor reflection on those who used the equipment last.

2.3.2 The steel band

For more accurate work, the steel band is preferred to the chain. As its name implies, it is made up of a steel strip, 13 mm wide, in lengths of 20, 25, 30, 50, and 100 metres. It has handles similar to those of the chain and is wound on a steel cross or drum for carrying purposes when not in use. For ease of use, the 20 m length is the most popular.

'Blued' steel bands may be graduated by brass studs at every 200 mm, with every tenth link marked by tabs or spots; while 'bright' steel bands may be fully graduated in metres and millimetres on one side and links on the other. Although not as robust as the chain, the band maintains its length with careful use and may be used as a standard measure. Care must be taken to see that it does not get kinked, and it should be dried and oiled after use.

Although the handles are usually included in the overall length, this is not always the case and this should be checked before starting work.

The effect of temperature change in terms of expansion and contraction of the chain or the band can generally be disregarded, since it is so minimal. Provided that the chain is correct at a normal temperature of 15.5 °C, the error is unlikely to exceed 1 in 4800 (approx.). Similarly, the slight elongation of the chain when tightened up, due to the elasticity of the metal of which the chain is made, is a very small fraction and may be ignored for ordinary work.

2.3.3 Arrows

Arrows, or pins, are made from steel wire, 300 mm or 375 mm long, with one end pointed and the other end formed into a ring for convenience in carrying. A strip of red cloth should be attached to the ring so that the arrow can be easily observed. A set of ten arrows is essential for marking off chain lengths as they are measured.

2.3.4 Ranging poles

Variously called ranging rods or pickets, these are wooden poles with a pointed steel shoe at one end for sticking in the ground. They are generally circular in section, 25 mm in diameter, although they may be octagonal, and painted in alternate bands of colour, i.e. red/white, red/black/white or black/white. The bands can be 200 mm or 500 mm wide and, although the colours are primarily for sighting purposes, the bands do allow the pole to be used for offset measurement.

For long sights over a distance of 800 m a white flag (a handkerchief or similar) will aid observation. Ten or twelve ranging poles will normally be sufficient for most surveys, but some adjustable three-legged stands should be available for use on hard surfaces where poles cannot be stuck in the ground. Poles may also be made in one piece or screwed together in sections, of aluminium or steel.

2.3.5 Whites

If the number of available ranging poles is limited, then poles already in the line may be taken out in order to allow the ranging-out to be continued. The poles thus removed are replaced by 'whites'. These are laths 600 mm to 1 m long carrying a piece of white cloth at the top. Gardeners' bamboo canes are ideal for this work. Whites may also be used to mark every ten chain lengths in a line.

2.3.6 Tapes

A tape is used for taking subsidiary measurements in the field. It is suitable for taking offsets, which are measurements taken from, and at right angles to, the chain line, or to fix adjacent points as on a boundary.

In past practice, all measurements taken with a tape, when used in conjunction with the Gunter's chain, would have been noted in links. Modern practice leans more to *all measurements being taken in the same units*, whether on the main lines, offsets, or ties, and it is therefore to the surveyor's preference whether links are used for the whole of the work as opposed to metres/millimetres.

Tapes are marked in metres, centimetres, and millimetres. The pure metric type will have links on the reverse face, as will the pure imperial type, but there are tapes available with metric measure on the main face and feet, inches, quarters, and eighths on the reverse.

Tapes are made in lengths of 10 m, 20 m, 30 m, 50 m, and 100 m. Unlike 'bands', which can be completely detached from their carrying frames and used like a chain, tapes wind into a hand-held carrying case made from leather or PVC-bonded steel. There is therefore only one end from which measurements are taken (see Fig. 2.5), the other end being attached to the central winding spindle of the case.

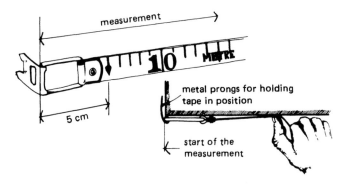

Fig. 2.5 Measuring using a tape. The user must ensure that the metal ring is included in the dimension.

Tapes are of two types: steel and fibreglass.

  i) *Steel* Graduated at every millimetre, the material may be 'blued' steel with etched markings, nickel-plated with black markings, or white-enamelled with black markings. Steel tapes are used in chain-survey work for important detail and short lines, while in building surveys (see Chapter 7) they are used where precise measurements are required, e.g. for steelwork centres etc. The tape should be cleaned and dried and wiped with an oily rag before being rewound into its case, or its life will be short.

ii) *Fibreglass*   Now in common use, this type is made of fibreglass encased in PVC. It is of the standard widths and lengths previously described and is white with graduations every 5 or 10 mm in black. Wetting has no ill-effect, and this type maintains its length with reasonable usage, although obviously it should be cleaned before being rewound into its case.

If the now obsolete linen tape is encountered, it should be treated with suspicion. This type of tape stretches very easily and, if used, will more than likely prove very unreliable.

When using a tape in the field, avoid unnecessary winding into its case. It is better to hook the tape around the little finger of one hand and let it hang in loose folds. This does not apply to the steel tape unless a glove is worn.

When winding in tapes, where the above is impracticable, hold the case in the left hand and allow the tape to be gently drawn through the first two fingers while turning the winder with the right hand. The first and second fingers of the left hand will bring off any dirt or water adhering to the tape and prevent their entry into the case, the principle being that it is easier to clean one's fingers than the inside of the case. At the end of the day, the tape should be drawn out of its case to its full extent, so that it can be completely cleaned before being rewound and stored away.

Finally, the end of the tape usually has a metal ring and hook at its beginning. Check whether this ring is included in the overall length before work begins (see Fig. 2.5).

2.3.7 The surveyor's folding boxwood rule

This is either fourfold, with four lengths joined by three hinges, or sixfold with six lengths joined by five hinges. The hinges swivel horizontally, and the unfolded length is 2 m graduated in millimetres on each side. The width is 25 mm. Colours available are natural wood with black markings or black with white markings.

2.3.8 The offset staff

This is a wooden rod, up to 3 m in length, often graduated in links on one side and normal measurements on the other. Unlike the tape, it can be handled by one person, although a long offset staff is unwieldy. It may be fitted with a hook at the top which is useful for lifting the chain over obstacles etc.

2.3.9 The plumb-bob

This is quite simply a weight held in free suspension so that the cord to which the weight is attached will always be vertical when there is no movement. The manufactured variety will invariably be a pear-shaped piece of gun-metal weighing about 450 g, with the lower end pointed and with a ring on the broad end to which the cord is attached. A plumb-bob may be improvised in the field with a stone and string or a nail and string, but for marking the *exact* spot under the point of suspension the manufactured bob is to be preferred.

The plumb-bob has several uses in surveying. It will be used when step-chaining over hilly ground and also as an attachment to a tripod when centring an instrument over an exact point on the ground.

2.3.10 Dropping-arrows

The dropping-arrow is a special arrow similar to a normal arrow except that it is heavily weighted at its point in the form of a plumb-bob. The dropping-arrow is used in step-chaining (see Section 2.4.6(b)).

2.4 Use of equipment for linear measurement

2.4.1 Chainmen

Two men are required for handling the chain. The *leader* pulls the chain forward and inserts an arrow at every chain length that is laid down. The *follower* directs the leader into line, tells him when to insert arrows, and collects the arrows as he himself arrives at them. (Note: In a two-man survey, the assistant would take the role of leader while the surveyor would act as follower.)

2.4.2 Handing over arrows

The leader starts off with ten arrows and at the laying down of the tenth chain length his supply will be exhausted. The eleventh chain length is then laid down and temporarily marked with a slip of wood or similar. The leader then walks back to the follower, who hands him the ten arrows that he has picked up as he followed down the line. The leader then substitutes an arrow for the piece of wood or temporary mark and the chaining continues as before.

2.4.3 Booking the tenth chain and marking the tenth chain

The surveyor should make a note in his field-book whenever ten chains have been measured. All arrows should be accounted for, and, to avoid the possibility of having to remeasure a long line, it is advisable to mark every tenth chain with a white.

2.4.4 Laying-down the chain

The leader, equipped with his ten arrows, drags the chain until he is brought
up by a gradual pull and directed into line by the follower. Once alignment is
effected, an arrow is inserted which marks the measurement of one chain
length. Care must be taken to ensure that arrows are inserted *vertically* and at
the side of the *vertical handle*, so that no error equal to the thickness of an
arrow or the thickness of the handle is introduced (see Fig. 2.6).

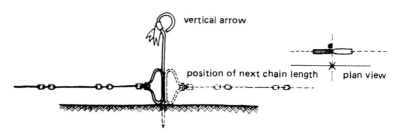

Fig. 2.6 Errors equal to the thickness of a marking arrow must not be introduced into
the chain-line measurement.

When the chain is in position, the surveyor walks along the line and, with
the help of an assistant, measures and notes all offsets and other measurements that need to be taken. When all measuring is completed, the leader then
drags the chain forwards and the laying-down process is repeated for the
second and subsequent chain lengths.

The starting point will be a predetermined survey station or a point on
another line where a check line is required to start, and the process of laying
down successive chain lengths and marking each one with an arrow is
continued until the end of the line is reached.

Some surveyors prefer to use two chains in concert, so that the handles of
successive chains come together as the line is extended. The merits of this
practice are somewhat devalued by the cumbersome business of having to
drag the first chain out of line and past the second chain (without disturbing
it) in order to align it in the 'third-chain' position.

Whichever method is adopted, care must be taken by the leader that the
arrows are not pulled out of the ground by a passing chain, and by the follower
when tightening the chain before the next arrow is inserted. If the ground is
hard and an arrow cannot be inserted, an arrowhead may be scratched or
chalked on the surface to indicate where the chain length ends, with the arrow
itself left on the ground pointing towards the mark.

These arrows are in turn picked up by the follower, who must also be careful
to hold his handle at the *side* of the arrow last inserted by the leader.

In order to achieve alignment, the leader usually carries a ranging pole so
that the follower can guide him into line. The chain itself is brought into line
by gentle snake-like ripples by the leader while the other end is held by the
follower.

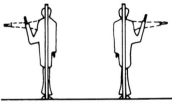

1. MOVE WELL OVER TO MY RIGHT (or LEFT) a rapid throwing-out type of movement with the hand.

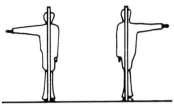

2. KEEP MOVING TO MY RIGHT (or LEFT) arm held out horizontally and kept in position until move is completed.

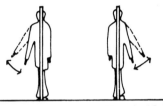

3. SMALL MOVEMENT TO MY RIGHT (or LEFT) slow sweeps of the arm to the side.

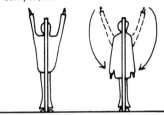

4. HOLD POLE STATIONARY

5. MARK! pole now on line.

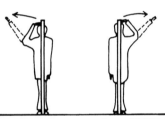

6. MAKE POLE PLUMB (i.e. VERTICAL) tap top of head and straighten arm in the required direction. This will indicate that only the *top* of the pole is to be tilted.

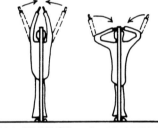

7. ALL FINISHED raise both arms and wave together then apart until signal seen.

8. COME TO ME arms raised high and open, then bring palms on to head–hold.

9. GO AWAY FROM ME the opposite signal to 8. Start with the palms on the head then move arms into high open position. Repeat until movement is completed.

All the above signals are stated as instructions FROM the surveyor. They are drawn as carried out by the surveyor and as SEEN by the assistant. All signals are essential since the distance between surveyor and assistant(s) may be as much as 100 m and the voice will be lost in the wind.

NOTE – Signals 6, 7, 8, 9 and 10 particularly can be adopted for levelling operations (see Chapter 4). Instructions will then be from the surveyor at the instrument to his staffman in the field.

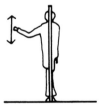

10. HURRY UP! extend right arm sideways but slightly bent. With the fist clenched, pump sharply up and down. Repeat as necessary.

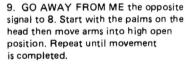

Fig. 2.7 Field signals

2.4.5 Ranging out

a) On flat ground   After the positions of the survey stations have been fixed, the straight lines between them are *ranged out* in order to facilitate chaining. Standing some distance behind the station pole, the surveyor directs his assistant into line and to insert the necessary number of ranging poles to form the line from the distant station. The line is *always* ranged out from the furthest station, towards the surveyor, since this gives greater accuracy. The assistant is guided by hand and arm signals as shown in Fig. 2.7.

Ranging poles must be erected *vertically*, and the surveyor observes the bottom of the pole first and then brings the upper part of the pole into line once the correct position of the foot has been determined. In order to give the surveyor clear vision, the assistant stands outside the line, holding the pole away from him at arm's length and between finger and thumb. He then moves into line as directed by the surveyor.

Although ten or twelve poles will normally be sufficient for most surveys, it should be borne in mind that on very undulating ground the poles must be spaced more closely than in open terrain. A sufficient number must therefore always be used to ensure that there is no departure from the true line when chaining. If there is deviation then all measurements taken at right angles to the line will be seriously affected, while to a lesser effect the length of the line will also be incorrect due to the deviation from the true line.

b) On hilly ground   Very often, due to undulations of some size, the last station point cannot be seen from the first, yet intermediate poles must be positioned for lining in the chainmen. The difficulty may be resolved by tying two poles together, although this is not very accurate or satisfactory. Two other methods may be adopted, as follows.

i) In Fig. 2.8, A and B are the two stations seen in plan, with the hill between them (as shown by the section). Two assistants with poles take up positions, one on each side of the hill, at C_1 and D_1 and facing each other so that the observer at C_1 can see the pole at station A and the observer at D_1 can see the pole at station B. By successively directing each other into line, their positions will be altered until finally they finish at C and D exactly on the line AB and then the poles are inserted.

ii) In Fig. 2.9, A and B are again the two stations with the hill intervening so that A cannot be seen from B and vice versa. A trial line (known as a random line) is set out from A with poles erected at C_1, D_1, etc. and will end at B_1 (unless by the greatest of good fortune the line ends on B, when there would be no problem). There is therefore an error at the end of the line amounting to BB_1, which is measured. AC_1, AD_1 and AB_1 are also measured. By application of the principle of similar triangles, it is found that triangle ADD_1 is similar to triangle ABB_1

$$\therefore \quad \frac{DD_1}{AD_1} = \frac{BB_1}{AB_1} \quad \text{or} \quad DD_1 = BB_1 \times \frac{AD_1}{AB_1}$$

Similarly the shift for any other pole is calculated.

Example   If $AB_1 = 400$ m, $AD_1 = 300$ m, and $BB_1 = 10$ m, then

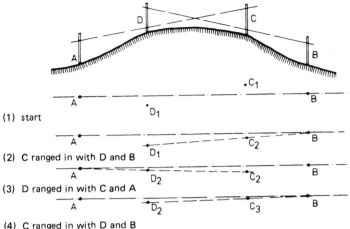

Fig. 2.8 Ranging out over a hill – method 1

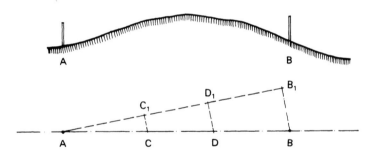

Fig. 2.9 Ranging out over a hill – method 2

$$DD_1 = \frac{10 \text{ m} \times 300 \text{ m}}{400 \text{ m}} = 7.500 \text{ m}$$

Once the shift distances for the intermediate poles have been calculated, it is a simple matter to bring C and D into line.

2.4.6 Horizontal measurement

In surveying, all distances are measured with reference to the *horizontal plane*. The drawn plan will then be a true projection of the ground.

a) Angle of inclination Land with a slope of 3° or less is usually considered level, since the difference between the length of the slope measurement and the horizontal measurement is negligible. This is illustrated in Fig. 2.10, where AC represents the surface of the ground making an angle α (the angle of

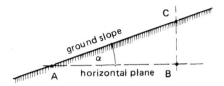

Fig 2.10 Horizontal measurement. Although the required horizontal distance AB is calculated, it is possible, where no great accuracy is involved, *to draw the section to scale* once the length AC and the angle α are known. The distance AB will then be scaled off from the drawing.

inclination) with the horizontal AB. If AC and α are known, it is a simple matter to calculate AB:

$$\frac{AB}{AC} = \cos \alpha \quad \text{or} \quad AB = AC \cos \alpha$$

i.e. horizontal length = slope length × cos (angle of inclination)

Where the angle is very small, the value of the cosine is nearly unity. Obviously the smaller the angle α, the nearer the length AC equates with AB. As the angle increases, the value of the cosine decreases rapidly; therefore, for large angles of inclination, the correction must be made or a serious error will be introduced.

Example 1   α = 1°, AC = 100 links, AB = 99.98 links.
Since the distances vary by only 0.02 link, (a matter of 4 mm) this difference can be ignored.

Example 2   α = 8°, AC = 100 links, AB = 99.03 links.
Now the distances vary by 0.97 link (a matter of 194 mm) which is far too great for the difference to be ignored and a slope correction would have to be made.

Slope correction and the measurement of α will be further discussed in Section 2.6; however, a simple method of measuring horizontal distance, without observation of the angle of inclination or use of calculations, is available as follows.

b) Stepping or step-chaining   The chainmen hold the chain, or part chain, *horizontal* from a point on the ground and into space. The next point of measurement on the ground is found by means of a plumb-bob or a dropping-arrow. The surveyor will determine, by eye, the horizontality of the chain (i.e. when it is at a right angle to the plumb-line) and an arrow is then inserted into the ground. This point will be the next position from which the chain will be held horizontally.

Figure 2.11 illustrates the method, which should *always* be carried out in a *downhill* direction – it would be impossible to hold B_1 exactly over B and resist a pull as the leader holds the chain at A. The method must therefore be as follows: handle at A; horizontal chain AB_1; plumb-bob to find B; handle moved to B; horizontal chain BC_1; plumb-bob to find C; and so on down the hill.

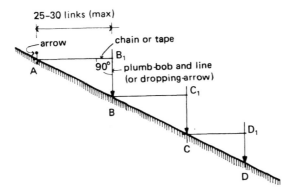

Fig. 2.11 Stepping or step-chaining

When chaining across a valley, it is necessary to erect a survey station, on line, in the trough and then *chain down both sides* to this point.

Care must be taken to keep chain lengths short – $\frac{1}{2}$ or $\frac{1}{4}$ chain – so that excessive sag does not result in positive errors in measurement.

c) Surface and catenary taping Where greater accuracy is required than the land chain will permit, other forms of measurement to obtain horizontal distance must be employed. However, these are outside the scope of this work and form a part of more advanced study, and it will suffice here to say that they exist.

2.4.7 Value of sloping ground

Hilly ground contains more superficial area than level ground but is of no more value. The natural growth of crops, trees, and plants is vertical, therefore no more corn or trees can be produced than on level ground. Buildings are also erected vertically, so by the same token sloping land will not allow a greater density of normal housing. Indeed, the value of sloping land may be considerably less than that of level ground because of the special problems it creates in its development. The cost of development of sloping ground – cutting and filling, special constructions, special drainage, etc. – must be weighed very carefully against the value and use of the end product so that an economic result may be achieved. Indeed, at survey stage, the question must be asked, 'Is development of this piece of land worthwhile?'

2.5 Instruments for 90° setting-out and slope measurement

2.5.1 The optical square

This consists essentially of two mirrors placed at an angle of 45° to each other and contained in a flat cylindrical metal box (Fig. 2.12). One of the mirrors, 'I', is called the *index glass* and is *wholly silvered*, while the other mirror, 'H', is

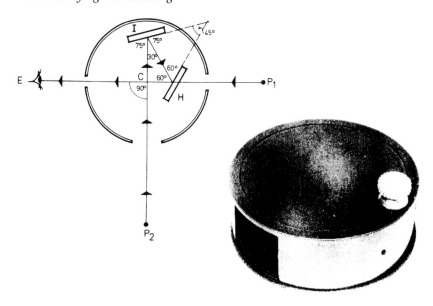

Fig. 2.12 An optical square

called the *horizon glass* and has its *lower half silvered* and its *upper half plain*. The horizon glass is set at 60° to the line of sight ECP$_1$, while the index glass is set at 75° to the line ICP$_2$.

The ray of light (image) from ranging pole P$_2$ is reflected from the index glass on to the horizon glass, and half the image of P$_2$ is then sent to the eye E. The eye also looks through the clear upper half of the horizon glass directly at P$_1$ (which would be on the chain line). When the pole P$_1$ and the half image of P$_2$ are seen as one pole then ECP$_1$ and ICP$_2$ are at right angles to each other. This is so because a ray of light reflected by two mirrors in succession is bent through an angle equal to twice the angle between the mirrors, i.e. through $2 \times 45° = 90°$.

Some types of square can be adjusted by means of a screw which rotates the horizon glass, while others need specialist attention. However, once properly set, no adjustment should be necessary unless the instrument has been subjected to gross abuse.

2.5.2 The prism (prism square)

A more modern version of the optical square is the prism or prism square (Fig. 2.13). The principle is the same but the reflecting surfaces are the internal faces of a pentagonal prism which are ground at an angle of 45° to each other. Having no movable parts, the instrument cannot get out of order.

Normally the prism square contains two prisms – one viewing to the right and the other to the left. This enables the observer to align himself exactly between two objects as well as setting off a right angle.

Fig. 2.13 A prism square

2.5.3 The cross-staff

This is a long-established instrument but is less precise than the optical square. Figure 2.14 shows three styles. In the form shown in Fig. 2.14(a) it consists of a brass box, octagonal in shape, about 75 mm deep and 60 mm broad. Down the centre of each face is cut a slot which contains a fine wire strained vertically. The staff can be fitted on the end of a ranging pole and be erected vertically. More properly, it should be attached to a tripod so that it can be placed exactly over a point on the ground by using a plumb-bob from the head of the tripod. By sighting through the slots, lines may be ranged off at 45° and 90°. For work of relatively low accuracy, it is quite a useful piece of equipment.

Figures 2.14(b) and (c) show types which consist of two pairs of vanes set at right angles to each other with a wide and a narrow slit in each vane. Again, each can be pole-or tripod-mounted for use at normal eye level, but they can only be used for ranging out lines at 90°.

2.5.4 The box sextant

Although little used in modern practice, and not entirely necessary in chain-survey work, this is nevertheless a neat, compact, and portable instrument by which angles up to 120° (or in some cases 140°) can be measured. It can be used as an optical square as well as for checking chain surveys by measuring some of the perimeter angles. It may also be used for fixing points of detail which are inaccessible (or nearly so), and to a limited degree it may be used in rough work as a substitute for a theodolite.

The disadvantages of the instrument are that it cannot measure an angle greater than 120° (or 140°) unless the angle is measured in two parts and also that, unless the objects observed are on the same horizontal plane as the

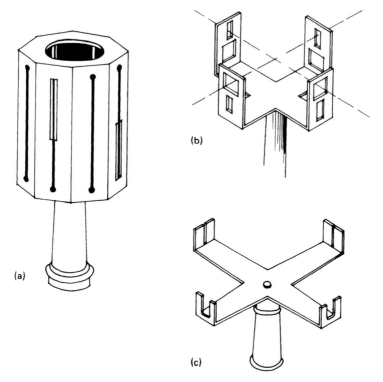

Fig. 2.14 Types of cross-staff. The one shown at (a) will also set out 45° angles.

observer, the angle measured is a slope angle and not the horizontal angle required in surveying technique.

The instrument will not be described here since adequate descriptions can be found elsewhere.

2.5.5 The clinometer

There are several forms of this instrument, which is used for measuring ground slopes. All are basically the same, and it is sufficient to describe the *Watkin's clinometer* for the principle to be understood. This consists of a counterweighted scale freely suspended so that the line EO is always hori zontal (see Fig. 2.15(a)), while the scale is divided from 0° in both elevation and depression.

Other types can have a graduated semicircle resembling a protractor with a light plumb-bob suspended from the centre (Fig. 2.15(b)), with AB (the straight-edge of the protractor) being used for sighting.

2.5.6 The Abney level

One of the most convenient instruments for measuring the angle of inclination is the Abney level, which consists essentially of a sighting tube, a spirit-level,

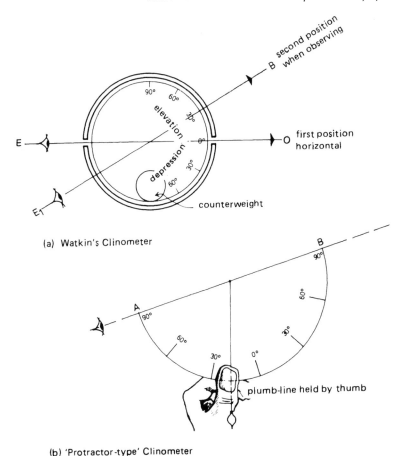

(a) Watkin's Clinometer

(b) 'Protractor-type' Clinometer

Fig. 2.15 The clinometer

and a scale graduated in degrees. This graduated scale is fixed rigidly at 90° to the pin-hole telescope (sighting tube) while a movable pointer, with the spirit-level attached, is fixed to the scale itself (see Fig 2.16).

A mirror fixed at an angle of 45° within the telescope, and covering half the width of the tube, enables the observer to see the uncased bubble as he looks

Fig. 2.16 The Abney level

through the telescope, a hole being cut in the top of the tube to admit the bubble image.

A vernier is often attached to the pointer which, like the graduated scale, is in bright steel. The tube case is black enamelled steel and the whole fits neatly into a leather velvet-lined case with a carrying strap.

2.6 Use of instruments for measuring angles

2.6.1 Setting-off 90° angles

Note: The chain may be used for setting out lines at angles to the main chain line, but this will be discussed in Chapter 3.

a) By optical square (maximum sighting distance 30 m)   In Fig. 2.17, a chain line is represented by P_3P_1. The surveyor stands on this line, holding the instrument exactly over the point C at which the right angle is required to be laid off. As in Fig. 2.12, he observes from E the pole P_1 on the chain line through the unsilvered portion of the horizon glass H. His assistant moves the pole P_2 until its image, reflected from the index glass I on to the silvered portion of glass H, is exactly below the pole P_1 seen by direct vision. The line CP_2 will then be perpendicular to the chain line. If the right angle is required to be set off on the other side of the chain line, the surveyor simply turns the instrument upside down and he and his assistant adopt the same procedure as described to locate the position of pole P_4.

Provided that either the chain line or the offset is horizontal, the optical square will set out a right angle on sloping ground.

Having located the position of P_2 by reference to P_1, if the instrument is turned upside down and P_3 is observed, P_2 should not need to be moved. If, however, the images do not coincide on the horizon glass, a second pole P_2^a must be placed by observation and direction. The true perpendicular to the

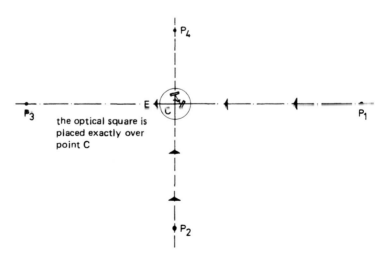

the optical square is placed exactly over point C

Fig. 2.17 Setting off a right angle using an optical square

chain line bisects the angle $P_2^a CP_2$. If a pole is placed at the true position of P_2, this can be used to adjust the optical square in the field (see Fig. 2.18).

b) By prism   The prism is held by the surveyor in his hand exactly over the point C, and the pole P_1 is sighted by direct vision *over* the prism. The rays of light forming the image of P_2 enter the prism from the side and are bent to the observer's eye. The assistant moves P_2 until the poles can be seen in vertical alignment, at which point the right angle has been set off.

c) By cross-staff   Unlike the optical square or the prism, the right angle is set out by direct observation of all poles. The cross-staff is placed on a tripod with a special receiving head, and the slots are made vertical. By use of a plumb-line, the cross-staff can be placed over the exact spot C in the chain line (Fig. 2.17). Poles P_1 and P_3 are observed through the slots as a test that the staff is on the chain line. When the surveyor is satisfied that the cross-staff is in line, he then observes through the slots at 90° to the line and when he can see pole P_2 through the appropriate slot this signifies that P_2C is perpendicular to the chain line.

The same technique is followed to set out a line at 45° to the chain line.

Provided that care is taken to ensure that the cross-staff is correctly over the chain line and observations are kept short, up to 20 to 30 m, a reasonable standard of accuracy should be achieved.

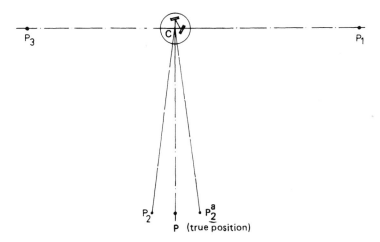

Fig. 2.18 Testing the optical square

2.6.2 Measuring slope angle

a) By clinometer   To measure the ground slope of a line AB, the surveyor stands at point A holding the Watkin's clinometer to his eye. The assistant stands at B with a pole having a clear marking which is the same height above the ground level at B as that of the surveyor's eye level at A. This mark is observed through ·the instrument and, if it is *higher* than the surveyor's eye level at A, the instrument will be tilted upwards (elevated). Since the scale is freely suspended, its position in relation to the horizontal will not alter and the angle of elevation will be read upon the scale from a fixed mark on the clinometer case (see Fig. 2.15).

Conversely, if the mark is *lower* than the surveyor's eye level at A, the instrument will be tilted downwards (depressed) and an angle of depression will be read from the scale.

Using the protractor type of clinometer, the surveyor observes the mark on the pole at B by sighting along the straight-edge while standing at point A. When the plumb-bob has taken up its position, the cord is fixed by the thumb and the angle is read off as in Fig. 2.15(b).

It must be appreciated that, for the above methods to be of use, the ground slope between A and B must be generally uniform, as illustrated in Fig. 2.19.

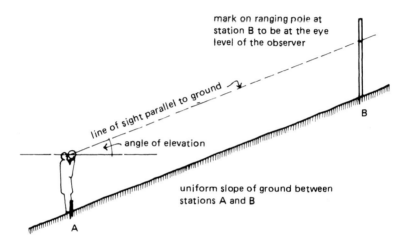

mark on ranging pole at station B to be at the eye level of the observer

line of sight parallel to ground

angle of elevation

uniform slope of ground between stations A and B

Fig. 2.19 Use of the clinometer

b) By Abney level The surveyor directs the instrument on to the mark on the pole, as for the clinometer method, at the same time adjusting the level by means of the levelling screw until he sees the bubble centred in the small mirror set at 45° inside the telescope. Since the index arm is fixed at 90° to the bubble tube and rotates with it on an axis at the centre of the semi-circular scale when the levelling screw is adjusted, the angle may be read off by means of the vernier scale on the index arm to the nearest ten minutes, secure in the knowledge that the line of sight will be parallel to the ground (Fig. 2.20).

c) Hypotenusal allowance By far the best method of achieving slope correction is to measure the ground on the slope, observe the inclination, and apply the correction at each chain length. This is referred to as the *hypotenusal allowance*.

In Fig. 2.21, AB is one chain length of horizontal distance required by surveying technique, α is the angle of inclination measured by instrument, AD is the distance of one chain laid on the ground slope, while DC is the correction required to ensure that point C is exactly above point B for plotting purposes. Thus

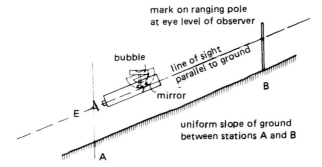

Fig. 2.20 Use of the Abney level

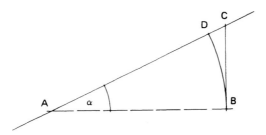

Fig. 2.21 Hypotenusal allowance

$$\frac{AC}{AB} = \sec \alpha \quad \text{or} \quad AC = AB \sec \alpha$$

Since AB is 100 links, the corresponding distance along the slope (AC) is 100 × sec α.

For example, if the angle of inclination is 10°,

$$AC = 100 \times \sec 10°$$

$$= 100 \times 1.015$$

$$= 101.5 \text{ links}$$

Hence the line AD needs to be extended by 1.5 links to ensure that point C is exactly over point B. Point D would be marked by a temporary arrow and the distance DC would be set off with a metal tape, after which the arrow would be moved to C to mark the corrected chain length. The subsequent chain lengths would each be corrected in the same way, any change in slope being observed by a clinometer or Abney level.

Any error in chaining will affect the accuracy of the result, as will any error in slope observation. This latter is of little consequence when the angle is small, but where the angle is large an error of $\frac{1}{2}°$ will affect the result considerably.

d) Alternative correction for slope When working in the field, it is often necessary to measure without regard to predetermined horizontal distances. In other words, whereas in **(c)** above, point C was found and made to lie above point B, the reverse may apply and, having arrived at point C, the position of point B and its distance from A may be calculated. The correction will be *deducted* from the measured slope length to arrive at the horizontal distance. In Fig. 2.22, the slope length AC is carefully measured, as is the slope angle α, and the error will be

$$AC - AB = AC (1 - \cos \alpha)$$

The function $(1 - \cos \alpha)$ is termed the *versed sine* of the angle α, and tables are available. The abbreviation for the versed sine is *versine*; hence the error can be written as $- AC$ versine α, the minus sign indicating that this correction must be deducted from AC to obtain AB. If versine tables are not available, cosine tables (or a calculator) may be used with the original formula.

This approach is preferable to the method stated in Section 2.4.6(a), where the length of AB was given by $AC \cos \alpha$. As a general rule, it is better to calculate a correction and either add or subtract this from the slope measurement, rather than modify the measurement directly.

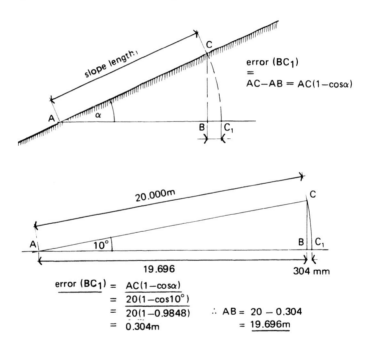

$$\text{error (BC}_1) = \underline{AC(1-\cos\alpha)}$$
$$= \underline{20(1-\cos10°)}$$
$$= \underline{20(1-0.9848)} \qquad \therefore AB = 20 - 0.304$$
$$= \underline{0.304m} \qquad\qquad = \underline{19.696m}$$

Fig. 2.22 Calculation of error as used for slope correction to find horizontal distance

Summary   It is most important that distances are measured *horizontally* so that the plan drawing is a true projection of the ground. Calculations and corrections for measurements taken on sloping ground are most important. It is quite useless to provide the architect, engineer, or quantity surveyor with a drawing which erroneously suggests that the area of land is larger than it in fact is. Buildings, roads, and other artefacts would be designed which would not fit on to the land, and prices and other calculations would bear no relation to the truth.

2.7 Equipment for noting information in the field

2.7.1 The field-book

This is required by the surveyor in order to record, by means of sketches and measurements, the work done in the field. The book is oblong in shape, 200 mm long and approximately 100 to 120 mm wide. It opens lengthways, being bound along one of the shorter sides. The book is used from the *back* and continued towards the *front*, which enables the surveyor to make his entries forwards and in the same direction in which he is walking. The pages of the book are of plain white paper with two parallel red lines, 12 mm apart, printed down the centre. These lines represent the chain and are used for *dimensions only*, and nothing should be drawn across them once measurement of a line has been started until the measurement is complete.

There are other types of book which are *not* to be recommended: those whose pages have only one single red line, or where horizontal pale blue lines are printed across the width of the page. These only confuse and detract from clarity and ultimate accuracy.

Figures 2.23 and 2.24 illustrate the use of the field-book.

The first entry is made at the bottom of the last page in the book and should be the date and place of the survey, together with the names of the persons involved with the work. Above this is placed a diagrammatic sketch of the pattern of triangles forming the framework, with all the survey stations being given a reference letter. On this diagram, each line can be given a number to indicate in which order it was surveyed, together with an arrowhead to indicate the direction in which it was measured. Compare the field-book page shown in Fig. 2.23 with the typical arrangement shown in Fig. 2.1.

2.7.2 Booking field notes

As the work proceeds, entries are made by the surveyor as measurements are taken by tape and chain. He also makes such sketches as are necessary, preparing the field notes in such a way that a draughtsman who was not involved with the fieldwork could produce the finished drawings.

Distances along the chain are noted between the double red lines which represent the chain, while offsets to points of detail are noted down immediately adjoining the figures giving the chainage at which they were taken (see Fig. 2.24). Any offset dimensions *must* be noted on the same side of the lines representing the chain in the field-book as they were taken in relation to the chain on the ground.

Measurements along a line may begin before the starting survey station is reached and continue beyond the end station. In this event, it is better to treat the first survey station as *zero* and note the distance to any boundary between the red lines but longitudinally. For progression beyond the end station, book the chainage as normal but highlight it as shown in Fig. 2.23 so that there is no doubt what is the chainage of the survey-station point.

Noting of lines should be begun and ended by the use of horizontal double lines at the first and last station. For example, '=== line AB' leaves no doubt that measurements were started at A and the surveyor walked to station B.

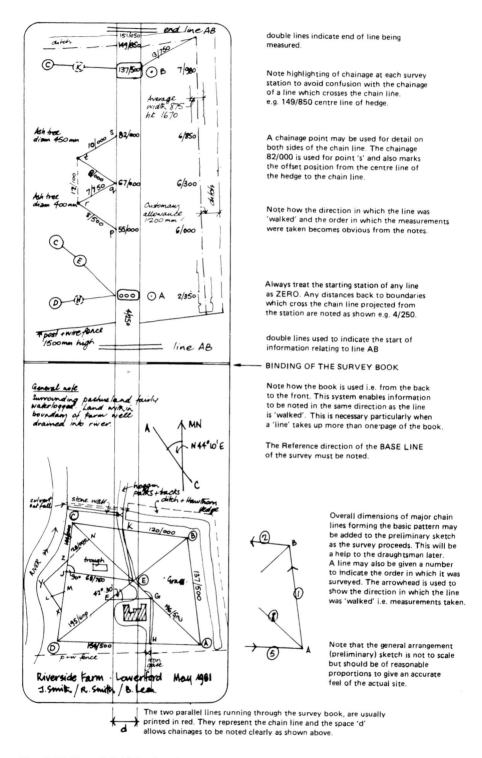

double lines indicate end of line being measured.

Note highlighting of chainage at each survey station to avoid confusion with the chainage of a line which crosses the chain line. e.g. 149/850 centre line of hedge.

A chainage point may be used for detail on both sides of the chain line. The chainage 82/000 is used for point 's' and also marks the offset position from the centre line of the hedge to the chain line.

Note how the direction in which the line was 'walked' and the order in which the measurements were taken becomes obvious from the notes.

Always treat the starting station of any line as ZERO. Any distances back to boundaries which cross the chain line projected from the station are noted as shown e.g. 4/250.

double lines used to indicate the start of information relating to line AB

BINDING OF THE SURVEY BOOK

Note how the book is used i.e. from the back to the front. This system enables information to be noted in the same direction as the line is 'walked'. This is necessary particularly when a 'line' takes up more than one page of the book.

The Reference direction of the BASE LINE of the survey must be noted.

Overall dimensions of major chain lines forming the basic pattern may be added to the preliminary sketch as the survey proceeds. This will be a help to the draughtsman later. A line may also be given a number to indicate the order in which it was surveyed. The arrowhead is used to show the direction in which the line was 'walked' i.e. measurements taken.

Note that the general arrangement (preliminary) sketch is not to scale but should be of reasonable proportions to give an accurate feel of the actual site.

The two parallel lines running through the survey book, are usually printed in red. They represent the chain line and the space 'd' allows chainages to be noted clearly as shown above.

Fig. 2.23 Use of field-book – 1

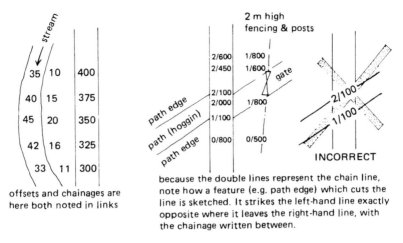

offsets and chainages are
here both noted in links

because the double lines represent the chain line,
note how a feature (e.g. path edge) which cuts the
line is sketched. It strikes the left-hand line exactly
opposite where it leaves the right-hand line, with
the chainage written between.

Fig. 2.24 Use of field-book – 2

Once the end of the line has been reached, a further pair of lines across the page will signify this fact. Except for lines such as check lines which may have little detail on them, a new page should be used for each chain line. Note the symbol used to denote the survey station and how it is kept away from the detail.

Direction lines, which indicate the various chain lines which start from or end at station points on the line being measured, should be shown clearly and in a way which does not conflict with the detail outlines. Note how this is done in Fig. 2.23.

Offsets   The sides of a road or river may be booked as shown in Fig. 2.24. All the offsets are measured *to* the chain line, the furthest being measured first and the nearest point last. When roads are laid out to specific widths, it may be sufficient to measure to the near side only and add a note stating the width of the roadway. In the case of rivers etc., always indicate the direction of flow.

Points of intersection with the chain line   Chain lines can cross each other, and other features such as fences, roads, ditches, etc. may also cross a chain line. When this occurs, the point or points of intersection must be shown, and direction lines must be sketched in the field-book. Note in Fig. 2.24 how the crossing angle is exaggerated and also how the detail line meets one red line and leaves the other. These two points are opposite each other with the chainage distance written between them. There should be no problem if the reader accepts that the red lines are in fact one line, pulled apart, in order that chainages can be noted between them, therefore any detail line can only cross the chain line at one point.

Hedges and ditches   All offsets taken to hedges should be taken to the centre of the root, and an indication should be given of the overall thickness of the growth or spread. Where a ditch is to be found behind a hedge, it is indicated by a line parallel to the hedge and 1200 mm from it, and would be drawn thus to scale. In the absence of other knowledge, it is assumed that the hedge and ditch are in the same ownership and that the hedge is planted on the earth taken out to form the ditch. The arbitrary distance of 1200 mm stated is known as the 'customary allowance', although it is as well to check the legal position before committing the survey to paper in drawn form (see Fig. 2.25).

Similarly, with fences it is customary to assume that the fence belongs to the person on whose land the supports are to be found, but again check title deeds to be absolutely sure.

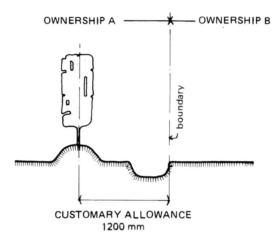

Fig. 2.25 Hedge and ditch

2.7.3 Pencils, eraser, penknife

All notes should be carried out in pencil. Ballpoint, pen and ink, and indelible pencil are not recommended for use in the field since notes could become totally unintelligible if they became wet. The use of an eraser may be necessary, although this should be minimised, while the need for a penknife is readily appreciated.

2.7.4 Summary

Within the broad outlines given, the booking of field notes is influenced by individuality. However, by getting into good habits from the outset, information will be clear and concise. Remember from Chapter 1 that, however desirable it may be for the surveyor in charge of the fieldwork, or one of his assistants, to complete the office task of plotting, there may be all manner of reasons which preclude this practice. Field notes should therefore be capable of translation into a scale drawing by any competent draughtsman, whether he has visited the site or not. An additional aid to this end is to use standard Ordnance Survey notation and British Standard conventions wherever possible on the field notes (see Chapter 3).

Exercises on Chapter 2

1. What are meant by the following terms: (a) preliminary inspection/reconnaissance, (b) survey station, (c) base line, (d) well-conditioned triangle, (e) survey line?

2. Upon what factors does the technique of chain survey depend?

3. Make a list with a short description of each item, of the instruments and equipment required in conducting a simple chain survey.

4. Sketch and describe the metric chain. How does the chain become altered in length? How can it be checked and adjusted?

5. a) A length of line was measured with a chain and found to be 1972 links. The true length of the line was proved to be 1986 links. What was the error in the chain? [+0.71 link]

   b) Using the above incorrect chain, an area was computed as 2.45 hectares. What was the true area? [2.485 hectares]

6. Explain how a chain line is laid down and how the procedure is extended for a total of *ten* chain lengths (including what happens between the tenth and eleventh chain measurements).

7. How would you range out a line AB on flat ground?
   If a hill intervened between A and B, from no point on which could A and B both be seen, explain with sketches how the line AB would be ranged out.

8. In Fig. 2.10, the angle of inclination α was measured as 9° and the slope length AC as 20.000 m. What is the horizontal distance AB? [19.754 m or 98.77 links]

9. a) Why is *horizontal measurement* adopted in chain-survey work?

   b) If a man contracts to plant trees 1 metre apart on a steep hill side at £x per hectare, should he be paid by horizontal measure or surface measure and why?

10. Compare the use of the cross-staff and the optical square for setting out right angles.

11. a) What is meant by hypotenusal allowance?

    b) In Fig. 2.21, the angle of inclination is measured as 8°. What is the distance DC so that point C will be exactly over point B? [0.195 m or 0.98 link]

3

Linear measurement – fieldwork practice

The broad principles of the chain-survey technique, outlined in Chapter 2, must now be extended to include the practicalities of the work, both in the field and in the office.

3.1 Errors in chaining

It has already been stated that, although nothing in surveying is entirely free from error, it is nevertheless possible to reduce error to an absolute minimum. In order to do this successfully, it is necessary *to know the likely sources of error* and what will be *their effect on the survey*. Armed with this understanding, effective precautions can be taken. The most likely errors in chain-survey work can be summarised as follows.

a) *Incorrect length of chain*, which will give either a *positive or a negative error*, depending upon the reasons for the incorrect length. The chain should be thoroughly inspected *before* being taken to site; be treated kindly rather than with disdain when being used; and be cleaned, inspected, and, if necessary, repaired *after* work is completed, so that it is left in good order for those who use it next.

b) *Chain not taut and properly aligned*, which will result in *a positive error*. The only precaution is to check the laying-down before taking measurements.

c) *Sag in the chain* will give a positive error. Remember to keep horizontal distances short when engaging in operations such as step-chaining.

d) *Expansion or contraction* due to temperature changes will result in either *a positive or a negative error*. This can be ignored in normal work, in a temperature of about 15°C.

e) *Elongation due to tensile stress* will give a negative error. Don't use the chain as though practising for a tug-of-war contest.

f) *Arrows improperly inserted* will give either *a positive or a negative error*. Arrows are fragile and easily bent, moved, or lost, so it is necessary to pay attention to the reasons underlying their use.

g) *Incorrect reading*, especially of fractional parts of the chain, will result in either *positive or negative errors*. Concentration is required, to the exclusion of idle chatter, when measurements are being taken.

h) *Insufficient allowance for slope* will give *a positive error*. Follow carefully the procedures outlined in Chapter 2.

i) *Incorrect booking of dimensions* will result in either *a positive or a negative error*. Only one remedy is possible: when a dimension is called out, note it down and then *call it back* as a check.

Chaining errors are mainly detected during the plotting process, when scaled check dimensions are compared with measurements obtained in the field. Errors found at this stage can be most inconvenient, especially if the site is a long way from the office. Doubtful triangles need to be rechained and, if costly and abortive work is to be avoided, the surveyor must take every care in the field to ensure that the possibility of error is kept to the minimum. He must always bear in mind what was said in Chapter 1 about errors, accuracy, and precision if the end results are to prove satisfactory.

3.2 Measuring offsets

Except in special cases, offsets are measured at *right angles* to the chain line, in order to determine the outline of boundaries or to fix other points necessary to the final plotting. The offset staff or the tape is used and, on occasion, the chain. Offsets should, however, be kept short – certainly no longer than 15 m in length – even when using the optical square to set out the right angle. Apart from the labour involved in measuring long lines, by keeping the offsets short there is less likelihood of error, whether due to the measurement not being taken at right angles or to incorrect length of the tape. In addition, even offsets shorter than 15 metres can produce plottable inaccuracies when drawn to scales of 1:200, 1:100 and larger.

With experience, the right angle can be judged by eye with an error no greater than 3° and usually better, and right angles judged by eye should be plotted by eye rather than by T-square and set square. Since the measurement is taken *from the point of detail to the chain line*, the tape may be pulled across the chain and then swung in an arc, the shortest arc which will just touch the chain giving the offset length. The point of which the chain is tangential to this arc will give the correct chainage for the offset. This is termed *a swing offset* and is illustrated in Fig. 3.1.

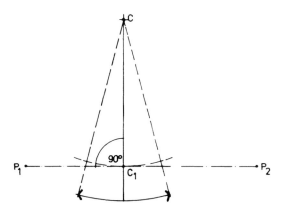

Fig. 3.1 Swing offset. The shortest measurement CC_1 which just touches the chain line P_1P_2 gives the *offset length* from C.

The number of offsets required will be determined by the nature of the boundary. If it is irregular, an offset will be taken at every point at which there is a break in continuity, along with all points of detail such as corners of buildings, gate posts, trees, poles, etc. Where there is a fair curve, offsets at regular intervals of, say, 5 or 10 m may be sufficient; while for a straight wall or fence, one at each end, with one at the middle as a check, can suffice.

There is a danger that, by placing too much reliance on things being straight or regularly curved, offsets will be taken at regular intervals and breaks in the continuity will be ignored or missed. Although enough offsets must be taken to define the boundary, the scale of the drawing should be borne in mind – it is a waste of valuable time to measure small deviations which cannot be shown on the plan, and useless dimensions can confuse the issue in many ways.

3.3 Division of the offset area

It can be seen from Fig. 3.2(a) that the *offset area* can be regarded as being divided into a number of trapeziums, each bounded by a part of the chain line (e.g. AD), a part of the boundary line (BC), and two offsets (BA and CD).

In taking the offsets, BA and CD are so spaced that BC can be regarded as a straight line, the interval AD depending on the nature of the boundary. By careful arrangement, the area of each trapezium can be calculated from the

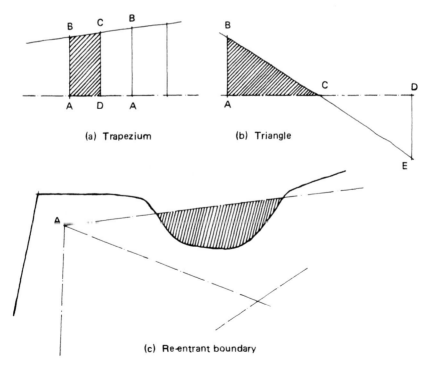

(a) Trapezium   (b) Triangle

(c) Re-entrant boundary

Fig. 3.2 Division of the offset area

formula *area of a trapezium equals half the sum of the two parallel sides multiplied by the perpendicular distance between them* so that the result is sufficiently accurate for all practical purposes.

e.g. area ABCD $= \dfrac{BA + CD}{2} \times AD$

Let the interval AD be 5 m and the offsets be BA = 3.22 m and CD = 3.25 m; then

$$\text{area ABCD} = \dfrac{3.22 \text{ m} + 3.25 \text{ m}}{2} \times 5 \text{ m}$$

$$= 16.175 \text{ m}^2$$

Where the boundary touches, or in fact crosses, the chain line, triangles will be created instead of trapeziums. In Fig. 3.2(b), the areas of the small triangles BAC and CDE may be calculated from the formula *area of a triangle equals half the base multiplied by the perpendicular height*.

e.g. area BAC $= \dfrac{AC}{2} \times BA$ and area CDE $= \dfrac{DC}{2} \times ED$

The whole of the offset area is then found by adding together the areas of all the small trapeziums and triangles. This total must be added to the areas of all the large triangles which divide up the space inside the chain lines (refer back to Fig. 2.1) in order to arrive at the area of the whole enclosure.

If at any point the boundary is re-entrant – i.e. the chain line is *outside* the boundary – the portion lying between the chain line and the boundary is regarded as a negative area and must be *deducted* (see Fig. 3.2(c)).

3.4 Fixing buildings and artefacts

The fixing of the positions of buildings may be achieved by means of *offsets, oblique offsets* (also known as *in-line ties* – see Fig 3.4–2), *perpendicular offsets, and ties.* It is also necessary to measure the building so that the plotting will be checked.

In Fig. 3.3, EA, FB, and GC are set off at right angles from the chain line P_1P_2, and their lengths are measured. Particular care must be taken to ensure that the offsets are *exactly at right angles*, otherwise *displacement* or distortion will occur which will increase as the length of the offset becomes longer.

A better method is shown in Fig. 3.4, where the lines of the walls AB, BC, and DC are projected through to meet the chain line at points E, F, and G respectively. The distances FB, EB, and GC are measured and noted down as *oblique offsets*. These fix not only the corners B and C of the building but also the direction of the lines of the walls BC, BA, and CD. The line EC is measured as a *tie line*. The chainage of points F, E, and G will be noted in the field-book between the centre pair of lines in the normal way.

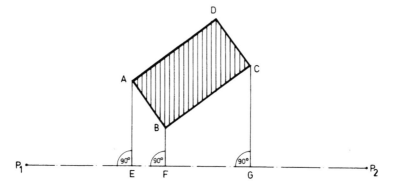

Fig. 3.3 Fixing buildings – 1

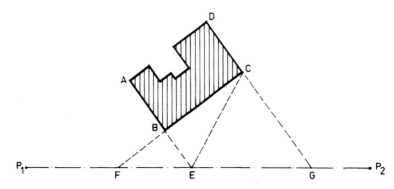

Fig. 3.4 Fixing buildings – 2

The building can, however, be fixed by a combination of perpendicular offsets and oblique offsets if this is more convenient on site. Any corner of a building, a tree, or some other points may be fixed by creating a small triangle to the chain line so that the plotting can be carried out by means of *intersecting arcs*, as explained in Chapter 1, Fig. 1.4.

When the position of a building has been definitely fixed in relation to the survey lines, the building itself should be surveyed as a separate item as described in Chapter 7.

3.5 Setting out a right angle by chain and/or tape

Several simple techniques can be used to set out a right angle by means of one or two chains, one or more tapes, or a combination of both. In all the following examples, the chain line is defined by the chain on the ground.

3.5.1 The theory and technique of the '3–4–5' triangle

From the *theorem of Pythagoras*, it is known that for a right-angled triangle *the square on the hypotenuse* (i.e. the longest side) *is equal to the sum of the squares on the other two sides.* For the triangle ABC in Fig. 3.5(a), the theorem will be written as

$$BC^2 = AB^2 + AC^2$$

and, if we let AB = 3 units, AC = 4 units, and BC = 5 units, then

$$5^2 = 3^2 + 4^2 \text{ (i.e. 25 = 25)}$$

proves the theorem for a triangle having sides in the proportion of 3, 4, and 5 units.

It must follow then, that if in the field a triangle is set up whose sides are in the proportions of 3 units, 4 units, and 5 units, one of the angles will be a right angle, and this angle will be opposite the longest side. The problem is to arrange for this right angle to be in the required place.

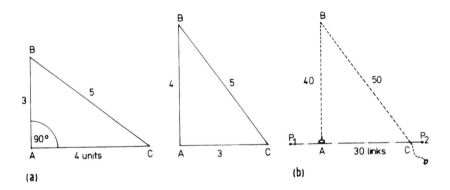

Fig. 3.5 The '3–4–5' triangle

In Fig. 3.5(b), it is required to set off from point A a line at right angles to the chain line P_1P_2. The method is as follows.

  i) Select the units to be used in the proportion of 3, 4, and 5; e.g. 30 links, 40 links, and 50 links.
 ii) Using a second chain, pin the handle on the chain at point A.
iii) At a point *30 links from A on the chain line* (point C) *pin the 90th link of the* second chain.
 iv) Holding the *40th link from A*, pull the chain taut (taking care not to pull the chain line out of alignment). Pin at point B.
  v) A right angle will be given at point A and the line AB will be perpendicular to the chain line P_1P_2.

The method works equally well if a tape is used instead of a second chain, or two tapes if AB is required to be longer than 40 links. Also, 3 m, 4 m, and 5 m lengths may be used in preference to links, while AC can be 4 units from A, with the chain or tape pulled taut by holding it at a distance of 3 units from A, as shown in Fig. 3.5(a).

3.5.2 From equal arcs

In Fig. 3.6(a), it is required to set off from point O a line at right angles to the chain line P_1P_2. The method is as follows.

   i) Mark two points A and B on the chain line, so that AO = OB.
   ii) Swing equal arcs from A and B to locate point C and mark its position.
   iii) Swing equal arcs from A and B to locate point D and mark its position.
   iv) The line joining C and D will pass through point O, and the line COD will be at right angles to the chain line P_1P_2.

The arcs may be swung by using tape or chain or by bringing two tapes together so that AC = CB and AD = BD.

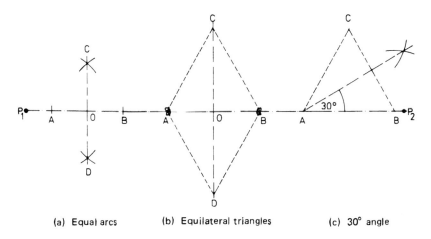

(a) Equal arcs   (b) Equilateral triangles   (c) 30° angle

Fig. 3.6 Setting out angles by chain

3.5.3 From equilateral triangles

In Fig. 3.6(b), it is required to set off from point O a line at right angles to the chain line P_1P_2. The method is as follows.

   i) Mark the two points A and B on the chain line so that AO = OB = 5 m and pin one handle of a second 20 m chain at A and the other handle at B.
   ii) Holding the *50th link*, pull this second chain taut without disturbing the alignment of the chain line P_1P_2. Mark point C.
   iii) Gently pull the whole second chain across the chain line and, *still holding the 50th link*, locate and mark point D.
   iv) The line joining C and D will pass through point O and be at right angles to the chain line P_1P_2.

3.6 To drop a perpendicular to a line from a point outside the line

There are several ways by which this can be done. The simplest and most suitable for the measurement of offsets is the *swing offset* which has already been explained (see Fig. 3.1, Section 3.2). The following methods may be used for more accurate work.

3.6.1 By equal arcs

In Fig. 3.7(a), let C be the point outside the chain line P_1P_2. The method is as follows.

    i) Swing a long arc from point C to cut the chain line in two places, A and B.

    ii) Mark these points and bisect the distance between them to locate point D.

    iii) The line from C through point D will be at right angles to the chain line, i.e. the required perpendicular.

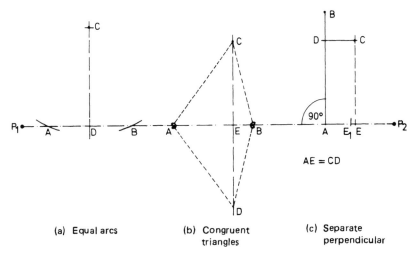

(a) Equal arcs   (b) Congruent triangles   (c) Separate perpendicular

Fig. 3.7 To drop a perpendicular from a point outside the chain line

3.6.2 By congruent triangles

The method is very similar to that described in Section 3.5.3. In Fig. 3.7(b), let C be the point outside the chain line P_1P_2. The method is as follows.

    i) Fix one handle of a second chain at point A on the chain line.

    ii) Run the chain to point C so that it is straight and taut.

    iii) Run the remainder of the chain from point C so that it is straight, taut, and has its handle touching the chain line at point B, thus creating the triangle ACB. Fix the handle at point B.

iv) Holding the link which coincides with point C, gently pull the whole second chain across the line and pull it taut to locate point D, thus creating the triangle ADB.

v) The line DC which cuts the chain line P_1P_2 at point E will be perpendicular to the chain line.

3.6.3 By a separate perpendicular

Where space is so restricted that the previous two methods are impracticable, the following method may be used – see Fig. 3.7(c).

i) Estimate by eye the point at which the perpendicular from C will meet the chain line P_1P_2. Let this be point E_1.

ii) A short distance from this estimated point E_1, at A on the line, erect a perpendicular and extend it to position B beyond point C.

iii) Measure the distance CD from C at right angles to the perpendicular AB. (Because the distance CD is small, the right angle required at D can be determined sufficiently accurately by eye.)

iv) Measure a distance, equal to CD, on the line from point A.

v) The point thus given will be the true point E where the perpendicular from C will meet the chain line P_1P_2.

The difference between E_1 and E should be small, and only a slight adjustment is likely to be necessary.

3.7 Setting out a 45° angle

In Fig. 3.8, it is required to set out a line from point B at 45° to the chain line P_1P_2. The method is as follows.

i) From point B, measure a suitable distance and mark point A.

ii) From point A, erect a perpendicular, using any of the methods previously described.

iii) Along this perpendicular, mark point C so that AC is equal to AB.

iv) The line joining C and B will be at 45° to the chain line P_1P_2.

An extension of this is to set up perpendiculars at both A and B and construct a *square*. The diagonals will give the desired 45° angle.

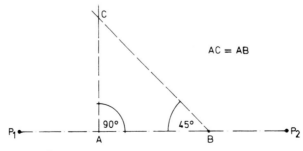

Fig. 3.8 Setting out 45°

3.8 Setting out 60° and 30° angles

Once the method of setting out a 60° angle has been achieved, it is simply a matter of halving it to set out 30°.

3.8.1 To set out 60°

The basic method has been partly described in Section 3.5.3 and Fig. 3.6(b). Let us suppose that the 60° angle is to be set out at point A in Fig. 3.6(b). The method would be as follows.

    i) Measure out the distance AB as 10 m and mark point B.
    ii) Fix the handles of a second chain, 20 m in length, at points A and B.
    iii) Hold the 50th link and pull the chain taut without disturbing the alignment of P_1P_2.
    iv) The triangle ACB thus created will be an *equilateral triangle* and all its angles will be 60°.

The alternative method is to use two tapes, one fixed at point A, the other at point B, and bring them together so that AB = BC = AC. Angles of 60° will again result because an equilateral triangle also has all its sides of equal length.

3.8.2 To set out 30°

Having carried out the operation as in Section 3.8.1 and set out an equilateral triangle, the method of obtaining 30° is as follows.

    i) Bisect the side of the triangle *opposite* to the required angle.
    ii) Join the bisecting point to the angle point and the original angle of 60° is thus divided into two angles of 30° (see Fig. 3.6(c)).

3.9 Obstructions in chaining

Although the preliminary reconnaissance and the pre-planning of the work to be undertaken should reduce this problem to a minimum, there are, nevertheless, occasions when the fieldwork is unavoidably obstructed and/or interrupted. It is essential that chaining is carried on in a straight line and that all distances along the line should be measured correctly from the starting point. The student should already be conversant with these requirements from the study of ranging out a line, as discussed in Chapter 2.

    Various obstructions can be met with – such as buildings, woods, hills, depressions, ponds, wide rivers, etc. – for which special methods can be adopted so that the line may be continued. Obstructions can be classified as follows:

a)   chaining free, vision obstructed – as in the case of rising ground;
b)   chaining obstructed, vision free – as when the line reaches a river, road, or pond;

c) both chaining and vision obstructed – as when the line meets a building or the edge of a wooded area.

3.10 Chaining free, vision obstructed

The most common occurrence of this situation is when the line has to be continued up and over rising ground, e.g. a small knoll or hill. The difficulty is simply to continue the ranging out and chaining in a straight line. This has been fully covered in Section 2.4.5. For very accurate work, the line should be continued by using an instrument such as a theodolite (see Section 6.12 for the method).

3.11 Chaining obstructed, vision free

Several solutions are available for this problem, making use of the chain and tape for setting out angles and erecting perpendiculars etc. as described in the earlier sections of this chapter, and also using optical instruments as described in Chapter 2.

3.11.1 Parallel diversion

This is a method whereby a line is laid out parallel to the part of the chain line which is obstructed. In the case of the pond shown in Fig. 3.9, there is no difficulty in ranging out the line beyond the obstruction, but physically laying down the chain to take measurements is impossible. The procedure then would be as follows.

i) Select two convenient points, A and B, on the chain line on either side of the pond. These are marked by the insertion of ranging poles.
ii) Erect perpendiculars AC and BD, at A and B respectively, with AC = BD and long enough to clear the pond. It is best to use the optical square for this work because the longer the perpendiculars, the more accurate the setting-off needs to be.

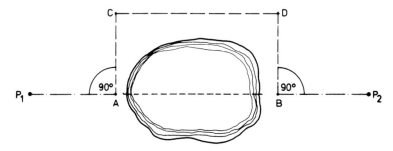

Fig. 3.9 Parallel diversion

iii) Insert poles at C and D.
iv) The line CD should be parallel to AB and equal in length.
v) The length of CD is measured and entered in the field-book and noted as *a parallel diversion*.

The lines AC, CD, and DB may also be used as *detail lines* to which offsets can be taken from the edge of the pond. A similar parallel diversion could be set up on the opposite side of the line to pick up the position of the pond's edge, if no other detail line was available.

3.11.2 Similar triangles

Figure 3.10 shows a bend in a river which interrupts the chain line P_1P_2. A method of dealing with this is as follows.

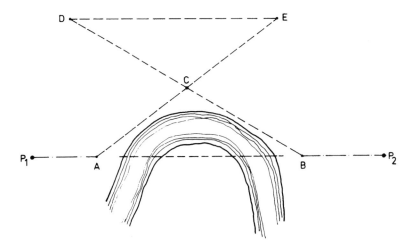

Fig. 3.10

i) Select points A and B on the chain line and as close as is practical to the river, marking each position with a ranging pole.
ii) Select a point C, so that the lines of sight AC and CB are clear of the river and give a well-conditioned triangle ACB.
iii) Continue line AC to point E, so that CE = AC; and similarly continue BC to point D, so that CD = BC.
iv) The triangle ECD thus created is equal in all respects to the triangle ACB, therefore DE is equal in length to the distance AB, which cannot be measured directly.

The details and measurements are entered into the field-book in the same way as any other detail related to the chain line. A note should be made that AB = DE.

Once again AC and CB may be used as detail lines, to pick up the boundaries of the river.

3.11.3 Right-angled triangle

In Fig. 3.11, the situation is exactly as in the previous section but the method now is as follows.

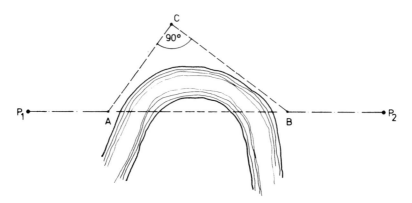

Fig. 3.11

i) A point A is chosen on one side of the river bend.
ii) A point C is selected so that line AC is clear of the bend.
iii) From point C a line CB, perpendicular to AC, is set off by means of the optical square, so that it is clear of the river.
iv) Where this line cuts the chain line, an arrow or ranging pole is inserted to mark point B. The lines AC and CB are measured and, from the theorem of Pythagoras,

$$AB^2 = AC^2 + CB^2$$

$$\therefore \quad AB = \sqrt{AC^2 + CB^2}$$

v) Again, the detail will be sketched into the field-book along with the dimensions *and* the above calculation of AB.

3.11.4 Other uses of triangles

Occasionally a line needs to be ranged beyond a *wide* obstruction such as a deep river or a busy road, where physical measurement of the line is impossible although access to the far side of the obstruction is available. This is known as the *classic case* and there are several solutions.

a) Chain line at, or nearly at, right angles to the obstruction   The line P_1P_2 is ranged across the river and poles A and B are placed in the line but on opposite banks (see Fig. 3.12(a)). The problem is to determine the distance between these poles so that the measurement of the line will be continuous.

i) Erect perpendiculars at A and C and fix points D and F so that AD = CF and DF is parallel to AC.
ii) Extend the line CF and locate point E by ranging in with points D and B. (Note: All the points will be marked by ranging poles.)

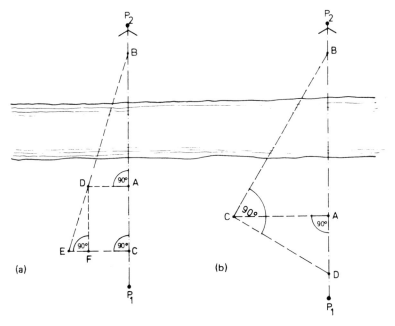

Fig. 3.12

Two *similar triangles*, EFD and DAB, have been created, from which the length of AB can be calculated:

$$\frac{AB}{FD} = \frac{AD}{FE}$$

$$\therefore \quad AB = \frac{AD \times FD}{FE}$$

This distance, together with sketches of the method and the calculations, is noted in the field-book and the survey can now proceed from point B on the far bank.

An alternative solution is shown in Fig. 3.12(b).

i) Points A and B are marked by poles in the line P_1P_2 but on the opposite sides of the river.

ii) Erect the perpendicular AC at point A, using the optical square.

iii) Select point C and at this point erect a perpendicular to CB which will cut P_1P_2 at D.

Triangles BAC and CAD are similar

$$\therefore \quad \frac{AB}{AC} = \frac{AC}{AD}$$

hence $AB = \dfrac{AC^2}{AD}$

By measuring the lengths of AC and AD, the required distance AB is readily obtained.

If a box sextant is available, the method is to fix points A and B on either side of the river as before and as shown in Fig. 3.13. A perpendicular is erected at A and a point C is selected so that AC is of sufficient length to give a well-conditioned triangle ABC. The angle ACB is observed with the box sextant, and the length of AC is measured.

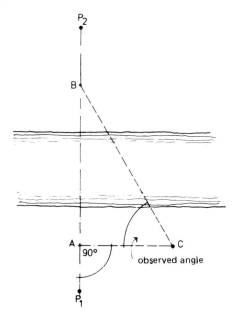

Fig. 3.13

$$\frac{AB}{AC} = \tan ACB$$

∴ AB = AC × tan ACB

It is a simple matter to calculate the distance AB, which is again entered in the field-book, together with the sketch and calculation, and the surveying is continued in the normal way.

b) Chain line crosses obstruction obliquely The oblique line P_1P_2 is ranged across the river and poles A and B are placed in the line but on opposite banks (see Fig. 3.14). Again the problem is to determine the distance between these poles.

   i) Range out the line P_3P_4 at an angle to the chain line P_1P_2. Let where the lines cross be point A.
   ii) Using the optical square, fix point C on the line P_3P_4 so that the angle ACB is a right angle.
   iii) Measure AC and locate point D so that AD = AC.
   iv) Using the optical square, erect a perpendicular at D to locate point E.
   v) Since the triangles EDA and BCA are equal, AB = AE.
      Measure AE and note this dimension in the field-book for AB.

Proceed with the survey from point B.

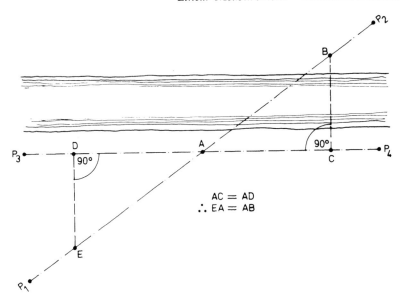

AC = AD
∴ EA = AB

Fig. 3.14

3.12 Chaining and vision both obstructed

This situation should not arise for main chain lines, but occasionally a detail line may be interrupted and it will be useful to know the method.

In Fig. 3.15, the line P_1P_2 is interrupted by a building and the usual method is to extend the principle of the *parallel diversion* (Section 3.11.1).

  i) Select two points A and B on the chain line as it approaches the obstruction.
  ii) Erect perpendiculars AE and BF equal in length.
  iii) The line EF is continued past the obstruction and two points on it, G and H, are chosen.
  iv) Erect perpendiculars GC and HD, equal in length to AE and BF.

Points C and D will be on the chain line P_1P_2 continued, and FG will be equal to BC. The length of the parallel diversion FG is measured and noted.

Great care must be taken to ensure that the perpendiculars are set off correctly and that their lengths are equal. Although the lengths AB and CD need not be equal, they should be at least three times the length of the perpendiculars in order to ensure that EH is parallel to the chain line.

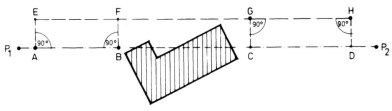

Fig. 3.15

3.13 Other aspects of dealing with obstructions

Although, in general terms, the above methods of dealing with obstructions will be found to be useful at times, it is advisable to avoid obstructions and interruptions wherever possible when setting the ranges of chain lines. This applies particularly in the case of the base line of the survey.

The student should also note that the methods described in Section 3.11.4 may be used to determine the position of an inaccessible point such as a tree on an island. A chain line would be run in line with the point and one of the methods applied.

3.14 Chain angles

When surveying an area such as a wood, a pond, a standing crop, etc., it will be impossible to measure diagonals in order to divide the shape into triangles. It is, however, necessary to fix the angles of the perimeter in some way, and this can be done by the measurement of *tie lines* using a chain or tape. Angles fixed by measurement in this way, without recourse to angle-measuring equipment, are referred to as *chain angles*.

Figure 3.16 shows the outline of an area where the measurement of diagonals is impossible. Survey stations are chosen at A, B, C, and D. The angles at the stations are fixed by *internal* tie lines, such as FE, or by *external* tie lines such as GH, KL, and MN.

In order to obtain external tie lines, one or both of the sides adjacent to the angle must be extended. In Fig. 3.16, for example, DC is extended to L and a

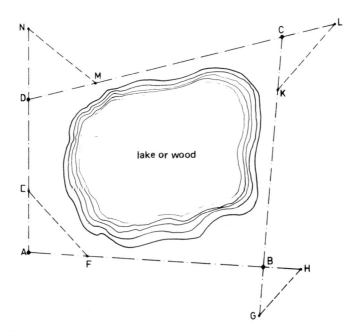

Fig. 3.16 Chain angles

point K on line BC is chosen. By measuring the line KL and knowing the distances CL and CK, the angle KCL is fixed.

Great care must be taken to secure good tie lines at angular corners. It is essential that they are long and give a well-conditioned triangle. Short lines render the work unreliable.

If the wood or pond etc. is so large or is so situated that a triangle cannot be run around it, it is probably much better to run a *theodolite traverse* (see Section 6.11), thereby measuring all angles by this instrument instead of using the chain.

3.15 Plotting the survey

There is no need to dwell at any great length on the subject of draughtsmanship. Numerous books are available on the subject, and even then there is only one way to become a competent draughtsman and that is to *practise*.

However, certain guide-lines can be given with regard to the plotting of a chain survey. Apart from the obvious skills of draughtsmanship, the person plotting the survey needs to have a thorough understanding of the principles on which the survey fieldwork was based and the reasons for which the information and drawings are required. It is better if the draughtsman has been involved in the fieldwork, although, with a good set of survey notes, any competent draughtsman should be able to plot the survey.

3.15.1 Equipment

This is basically as listed in Section 7.2,,with the following additions and notes.

a) **Long steel straight-edge** to rule the long chain lines which form the triangulation pattern.
b) **Parallel rule** useful for transferring lines and points across the drawing paper.
c) **Protractor (360°)** for plotting horizontal bearings. The author's is a large, 10 inch (254 mm), brass circle which formerly belonged to his grandfather and the larger the protractor the more accurate the plotting. The majority of survey points read from the instrument (Theodolite etc.,) will be outside the extent of all but the largest protractors and substantial errors can be produced by extrapolating the inaccuracies of a small-diameter protractor over a 'long' distance. When using a 'modern' plastics protractor care must be taken to ensure that, because of its light weight, it does not move about when in use.
d) **French curves** useful when plotting highway boundaries, kerbs, and contours.
e) **Railway curves** for plotting roads and railways.
f) **Offset scales** used to facilitate the plotting of offsets. The scales are 50 mm long, being divided in exactly the same way as the long scale with which they are used. The long scale is laid down on the drawing, parallel to the chain line, and the offset scale is placed at right angles (see Fig. 3.17). The offset scale is then moved along the chain line to the various chainages

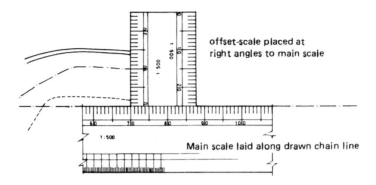

Fig. 3.17 Use of offset scales

and the appropriate offset is marked off at each position. Some offset scales have the zero marked at the centre of the scale to enable offsets on both sides of a line to be plotted.

g) **Steel pricker** for marking station points and chainages, in preference to a pencil dot.

h) **Paperweights**   Blocks of lead covered with leather are now rather difficult to obtain, although most 'older' established offices will doubtless have a supply. Glass blocks, often bearing trade or company advertising, also serve the same purpose.

It is much better to work on a drawing held by paperweights when plotting offsets by means of a T-square and a set square. Each chain line in turn requires to be set parallel to the blade of the T-square so that the offsets can be drawn at right angles to it with the aid of a set square. The weights allow the drawing to be turned much more readily than if it is fixed to the drawing-board by means of drawing pins or draughting tape.

i) **Spring clips**, which slide or clip over the edge of the drawing-board, can also be used in preference to pins (see also Section 9.3).

3.15.2 The plotting

a) The framework   Firstly the scale to be used must be decided upon so that the approximate size of the finished plan can be calculated. A sheet of paper can then be used which is large enough to take the whole of the work. Obviously, for very large areas, more than one sheet may be required. The student should always remember that the design of the layout is very important. The plan should be positioned so as to give the sheet a balanced layout.

The plotting is carried out in exactly the same sequence as the measurements in the field. Using the steel straight-edge, the base line of the survey is set out and the station positions at each end and also any others within the length are marked by means of the steel pricker rather than by pencil. All station marks should be circled and given the identification letter used in the field. All the triangles are plotted off the base line using the beam compasses; and all the station points are marked, circled, and lettered.

The student should note that compasses are never set against a scale rule – a *measuring line* is drawn at the bottom of the sheet, out of the way of the drawn work, and all arc lengths are set out to scale along this line.

When the triangulation pattern has been plotted by the intersection of arcs, the lengths of the *check lines* should be scaled off the drawing and compared with the field-note dimensions. Where errors are found, the plotting should be checked and then the field notes. If the reason for the discrepancy cannot be found, a return visit to the site may be necessary in order to re-measure the relevant parts.

Once the whole of the framework has been plotted *and proved*, the triangulation pattern should be inked in using *thin ink lines* of carmine, sepia, or cobalt-blue colour. The *detail* may now be plotted.

b) The detail   Again, the detail should be plotted in the same order in which the measurements were taken in the field. Successive chain lines should be *completed in turn* until all the plotting is finished.

c) The offsets   Normally, if right angles were judged by eye in the field, they are similarly plotted by eye. If the optical square was used in the field, plot by means of offset scales or T-square and set square.

d) Ties   Chainages are pricked off in the normal way, then, using compasses set to the scale distance required, arcs are swung to intersect at the point of detail.

3.15.3 Completing the plan

Once the plotting has been completed in fine pencil lines, the details are drawn in carefully. This is termed *lining-in*. Hatching and cross-hatching may be used for buildings and greenhouses respectively, and conventional symbols, as shown in BS 1192, 'Building drawing practice', should be added as appropriate (see Fig. 3.18). Offset points should not be simply joined by straight lines unless the boundary between them *is* a straight line. The student must remember that the drawing is a representation of the actual site; therefore, the *selection of detail points on site should be made with the plotting in mind* if a true representation is to be achieved. The lining in is done in *black ink*. Extremely good draughtsmen can achieve excellent results with pencil only and, although this is less permanent than ink, it can be acceptable in certain cases.

The drawing should be 'titled up' and additional information in the form of date . . ., surveyed by . . ., drawn by . . ., scale . . ., client's name . . ., office name . . ., job number . . ., sheet number . . ., etc. should be printed on the sheet in the bottom right-hand corner. A full title should be printed along the bottom edge of the sheet, to facilitate finding the drawing when it is stored in a plan chest – sheets titled at the top require the searcher to remove all the drawings from the chest drawer, often with unfortunate results. Although the scale must be written on the sheet, a *drawn scale*, usually the length of the longest line of the survey, is an advantage. Drawings are subject to movement through temperature change and a drawn scale will always remain consistent with the drawing. Borders or margins should *not* be put on drawings merely to

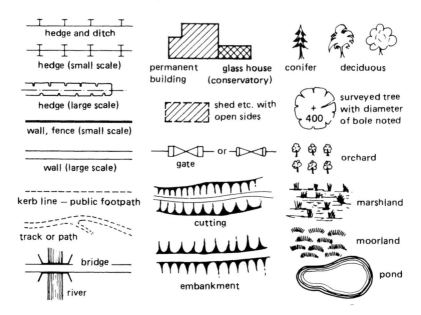

Fig. 3.18 Some conventional symbols used on maps and plans. The student should also refer to BS 1192 and O.S. sheets.

'enhance' them – it is better to let the drawing speak for itself within the confines of the overall sheet size.

Finally, a *north point* is essential and this can be parallel to the side of the sheet, i.e. north is at the top of the sheet, although this is not obligatory.

An example of a *finished* survey drawing is shown in Fig. 3.19.

3.16 Construction of scales

Having read Section 1.5.7, the student will appreciate that drawings are not usually made to the same size as the object which they represent, except in special circumstances, e.g. full-size detail (see Chapter 7). Drawings are therefore drawn to various scales, suitable for the task, and it is the *representative fraction* of the scale used which indicates the proportion that each line on the drawing bears to the object delineated. The scale of a drawing may be stated

a) in *words*, e.g. 20 mm to 1 metre, etc.
b) by its *representative fraction*, e.g. 1/20, 1/50, etc.
c) by a *ratio*, e.g. 1:20, 1:50, etc.
d) by *drawing a line divided into equal parts*, each representing the unit used.

When a scale shows equal divisions only, it is termed a *plain scale*. Generally, for scales to be of use they must

   i)   be divided with great *accuracy* and be carefully numbered or *figured*,
  ii) be long enough to measure the principal lines of the drawing,
 iii) have the name of the scale, the representative fraction or the ratio, and the numbering of the primary divisions shown.

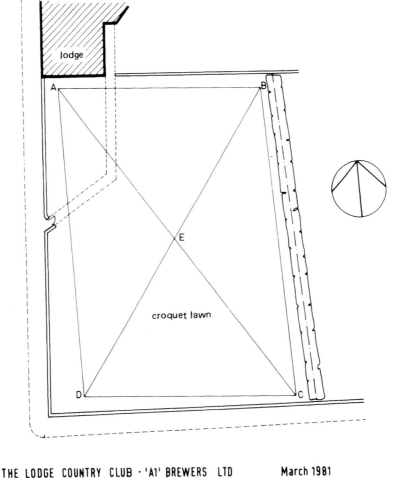

Fig. 3.19 Survey drawing

Metric scales are *close-divided* – which is to say that every major division is fully subdivided – and commercially obtainable scales are read *from left to right*, i.e. zero is at the left-hand end of the scale (see Fig. 3.20).

3.16.1 To subdivide a line of given length into *n* equal parts

The principle of subdivision of long lengths, rather than the repetition of short lengths, is of the utmost importance in accurate draughtsmanship and the construction of scales.

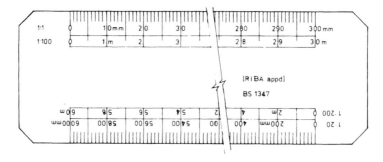

Fig. 3.20 Metric scales

In Fig. 3.21, AB is a line of given length which is to be divided into *n* equal parts. The method is as follows.

i) Draw the line AB to the required length.
ii) From A, set up the line AC at any convenient angle to AB. (Since the accurate determination of the exact point of intersection of two lines having appreciable thickness is difficult when the angle of intersection is acute, the angle BAC should not be made too acute.)
iii) Lay off along AC the required number of equal divisions, 1,2,3, . . ., *n*, by means of either dividers or a ruler showing known equal divisions.
iv) Join *n*B and, from points 1,2,3, . . ., (*n* − 1), draw lines parallel to *n*B to cut AB at 1',2',3',. . .,(*n* − 1)', following which AB will be divided into *n* equal parts.

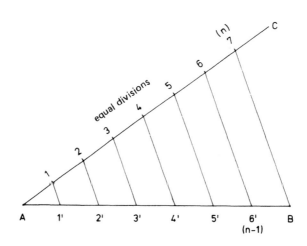

Fig. 3.21 Division of a line into *n* equal parts

The proof of this construction is based on the fact that, in the triangle AB*n*, parallels to the side *n*B divide the sides A*n* and AB into parts proportional to one another.

3.16.2 To construct scales

The construction of a scale is a simple operation provided the principle of subdividing a line into a given number of equal parts is thoroughly understood.

Example To construct a scale of 1:50 to read 5 metres.

The scale could be written as 20 mm to 1 m; therefore the length of the scale will be 5 × 20 mm = 100 mm (1 decimetre) long.

   i) Draw two parallel lines 100 mm long and 3 mm apart.
  ii) Divide them into *five* equal parts and subdivide *each space* into *ten* equal parts.
      Each major division will be 20 mm long and will represent 1 metre on the ground. Each subdivision will represent a length of 100 mm.
Most commercially obtainable metric scale rules have each subdivision further divided into *two* equal parts so that a 1:50 scale can be used to read 50 mm and 100 mm as well as whole metres.

Generally, metric scales are fully divided into 1 mm subdivisions, with each 1 mm division reading a different amount depending on the scale of which it forms a part.

3.17 Calculation of areas

A sizeable part of the work of an architect or surveyor is concerned with the computation of areas. The reasons for calculating the area of land or space within, or occupied by, buildings are many and varied and some are dealt with in Section 7.9.

For the purposes of calculation, areas on plans and drawings may be classified as

   i) *Straight sided* which are capable of either giving a result by the application of a specific formula or being subdivided into parts to which specific formula can be applied, e.g. triangles, rectangles, etc.
  ii) *Regular shaped but not straight sided*, e.g. circles, arcs, etc., having their own specific form of calculation method, and
 iii) *Irregular sided* which require special methods of calculation to be adopted.

The fastest and most accurate method of obtaining an area from a drawing is to use a planimeter (see Fig. 3.22), although the use of formulae and specific methods of calculation can give an acceptable accuracy dependent upon the task in hand. Computer applications are dealt with in Chapter 10.

3.17.1 The planimeter

The measurement of areas by the *planimeter* is the most efficient and fastest method for small or highly irregular shaped figures. The use of the instrument

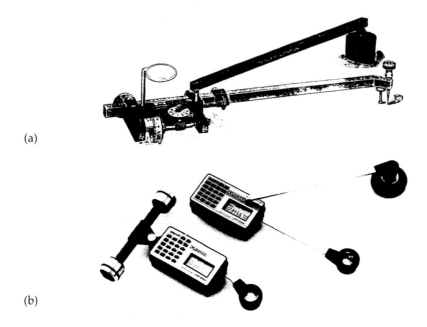

Fig. 3.22 Planimeters: (a) Traditional (Stanley) type, (b) Modern types – 'Sokkisha' KP80N polar planimeter and KP90N roller planimeter

is termed *mechanical integration* and the expression '*calculating the area under the curve of a function*' is also used by some surveyors because of the vernier scale and drum system of recording the measurement.

The ordinary planimeter (see Fig. 3.22) consists of:

   i) *a pole block* for anchoring the instrument to the surface of the drawing, not only by its weight but also by means of *a fine steel retaining needle.*
  ii) *the pole arm*, of which one end is pivoted at the pole block and the other end is pivoted at *the integrating (i.e. measuring) unit.*
 iii) *the tracing arm* which is attached to the measuring unit at one end and has *a tracing point* at the other end. The length of the arm may be fixed or variable (i.e. capable of being extended by means of a sliding bar.).
  iv) *the integrating unit (or measuring unit)* which consists of an integrating wheel (or disc) which runs over the paper and operates a drum divided into 100 parts. Readings to 1/1000th of a revolution are read by *vernier scale* (see Section 6.6.1) or estimated or read by digital readout. Whole revolutions are recorded on another indicator. The planimeter can be used in two ways
   (a) with the pole *outside* the figure being measured, which is the most usual method of working, or
   (b) with the pole *inside* the figure being measured.

a) with the pole OUTSIDE the figure the procedure is:

   i) ensure that the drawing is on a flat horizontal surface;
  ii) place the pole block *outside* the area to be measured, press the point of the retaining needle into the paper, ensuring that the tracing point can reach any point on the outline or boundary of the figure;

iii) place the tracing point exactly on a mark on the outline of the area being measured;

iv) read, and note down, the number of revolutions (to three places of decimals) shown on the drum and vernier;

v) carefully move the tracing point *clockwise* around the outline or boundary, terminating its run exactly on the starting point;

vi) read, and note down, the new vernier readings;

vii) calculate the difference between the two sets of readings which will be the number of revolutions made by the disc in circumscribing the figure;

viii) repeat stages (ii) to (vii) until *three* consistent values are obtained and *mean* these for a working result.

The instrument has a scale factor which is normally found on a plate fixed to the inside of the box lid, along with other information. With the *sliding bar instrument* the scale factor varies with the position of the tracing arm at the integrating unit, the arm being graduated with settings for a variety of drawing scales in order to give results at plan scales when the factor is applied. A fixed bar planimeter is generally constructed so that the number of revolutions gives the area measured in square millimetres/centimetres or square inches direct.

Example: Calculate the area of a piece of land which has a plan area of $1550mm^2$ as measured by a *fixed-arm planimeter* if the scale of the plan is 1:1250.

From Section 1.5.7, it will be appreciated that at a scale of 1:1250, $1mm^2$ represents $1250mm^2$,

$$\therefore \quad 1550mm^2 = 1550 \times 1250 \times 1250mm^2$$
which is $2421875000mm^2$
i.e. 0.242 hectares or thereabouts.

Once the scale factor of the instrument is known, the area A of the figure will be given by

$$A = scale\ factor \times number\ of\ revolutions$$

b) **with the pole INSIDE the figure** the procedure is generally as in a) but *a constant*, engraved on the tracing arm, must be added to the difference in readings each time the area is 'swept out'.

The constant represents the area of *zero circle of the planimeter*, i.e. of that circle which will be swept out when the plane of the integrating disc is such that it is lying through the pole and the disc does not revolve.

c) **generally** the operation is very rapid although the quality and accuracy of the results depend upon the level of skill of the operator. If the area to be measured cannot be circumscribed from one pole position, it can be divided into smaller areas (by means of thin pencil lines) each area being measured separately and the results added together.

3.17.2 Areas of straight sided figures (see Fig. 3.23)

a) **The triangle** which is *a plane figure bounded by three straight lines*, is the simplest and the most useful of figures used in surveying practice. If the

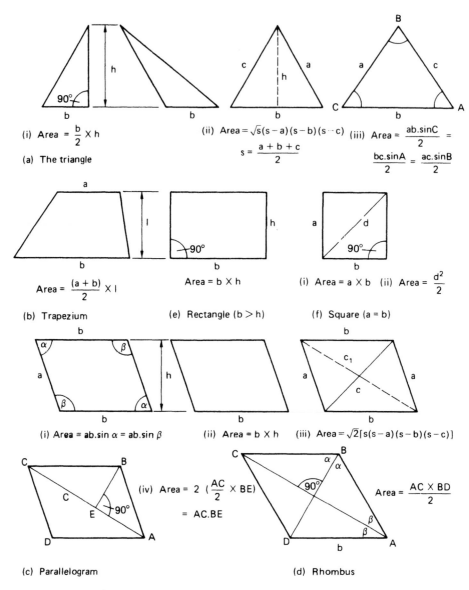

Fig. 3.23 Areas of straight-sided figures

base length b and the vertical height h of any plane triangle are known the area A can be calculated from the formula

$$A = \frac{b}{2} \times h$$

Alternatively, if the sides a, b and c of the triangle are known, the area A can be obtained from

$$A = \sqrt{s(s-a)(s-b)(s-c)}$$

where s = *half the sum of the sides* ($s = \dfrac{a+b+c}{2}$)

However, if two sides and the *included angle* are known, the area A can be obtained from

$$A = \frac{ab.\ \sin c}{2} = \frac{ac.\ \sin B}{2} = \frac{bc.\ \sin A}{2}$$

b) **The trapezium (or trapezoid)** which is a *quadrilateral having one pair of opposite sides parallel*, is also extremely useful as an aid to the calculation of areas (see Fig. 3.2). One application is dealt with in Section 3.3. From Fig. 3.23 it can be seen that the appropriate formula is

$$A = \frac{(a+b)}{2}\ \ell$$

where a and b are the parallel sides, ℓ the perpendicular distance between them and A the area being calculated.

c) **The parallelogram** is *a plane four-sided figure whose opposite sides are parallel.*

 i) If the height is not known but the sides and any angle are known, the formula with which to calculate the area A is

$$A = ab \sin \alpha$$

 Note: $\sin \alpha = \sin \beta$ because the angles are supplementary

 ii) if the base b and the height h are known then the area A is found from

$$A = b \times h$$

 iii) If two adjacent sides a and b and a diagonal c are known the parallelogram is divided into two *similar and equal triangles* by the diagonal. The area A of either triangle can be found by the formula

$$A = \sqrt{s(s-a)(s-b)(s-c)}$$

 and the area of the parallelogram = $2A$

 iv) If the diagonal and the perpendicular on it from an opposite vertex are known then the area is the product of the diagonal and the perpendicular.

d) **A rhombus** is *a parallelogram whose sides are all equal*

 i) If the *two* diagonals are known, the area A is *equal to half the product of the diagonals.*

 In addition the diagonals bisect

 i) each other at right angles,
 ii) the angles they connect.

e) **A rectangle** is *a parallelogram which has ONE of its angles a right angle* from which it follows that all four internal angles are right angles. The area A is the product of the base and the height.

f) **A square** is *a rectangle which has TWO ADJACENT sides equal* from which it follows that all its sides are equal and all four internal angles are right angles.

i) If the lengths of side are known then area A is *the product of adjacent sides.*

ii) If only the diagonal is known then area A is found from

$$A = \frac{d^2}{2} \text{ where } d \text{ is the diagonal since}$$

$$d = a\sqrt{2} \text{ where } a \text{ is one side.}$$

g) **Any rectangular figure** can be divided into triangles and quadrilaterals by drawing straight lines in the form of diagonals and perpendiculars (see Figs. 2.1, 7.6 and Section 7.4.3). The areas are easily found and when added together give the area of the whole figure.

3.17.3 Areas of regular shaped figures not having straight lines (see Fig. 3.24)

a) **A circle** *is a plane figure bounded by a curved line, every point on which is at the same distance from a certain fixed point, i.e. the centre, in the figure.*
   i) The area A is given by $A = \pi r^2$
   ii) The circumference c is given by $c = 2\pi r$ or $c = \pi d$

It is not necessary to prove these mathematically at this point although the student should understand that π is a ratio equal to the circumference c divided by the diameter d

$$\pi = \frac{c}{d} = \frac{22}{7} = 3.1416 \text{ (3.14159)}$$

b) **An annulus** *(or circular ring) is an area enclosed between two concentric circles,* with the radii of the circles being R and r respectively, the area A is given by
$$A = \pi R^2 - \pi r^2 = \pi(R^2 - r^2) = \pi (R + r)(R - r)$$

c) **An arc of a circle** *is any portion of its circumference* the length ℓ of which is given by $\ell = \frac{\pi r\theta}{180°}$ when θ is the angle subtended at the centre of the circle.

It is necessary to be able to calculate ℓ in order to obtain the area of a sector.
d) **A sector of a circle** *is the figure bounded by an arc and the two radii drawn to its extremities.*
The area A is given by $A = \frac{1}{2}\text{arc} \times \text{radius}$. If the student thinks of the sector as *a sort of triangle* on a curved base, the height at every part being the radius, r, and the base being the *arc, a,* then there should be no difficulty in remembering the formula. From c) above, $\ell = \frac{\pi r\theta}{180°}$ therefore the area A

of the sector can be written

$$A = \frac{1}{2}\frac{(\pi r\theta)}{180°} \times r = \frac{\pi r^2\theta}{360°}$$

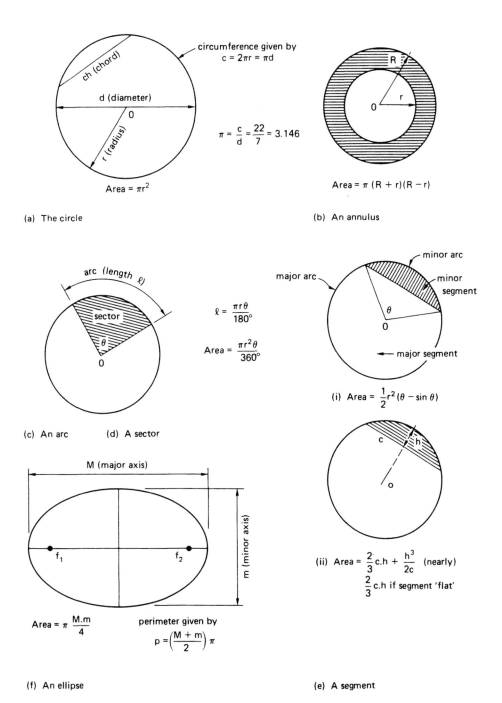

(a) The circle

(b) An annulus

(c) An arc   (d) A sector

(f) An ellipse

(e) A segment

Fig. 3.24 Areas of regular shaped figures not having straight lines

e) **A segment of a circle** *is the figure bounded by an arc and its chord.* The student should note that a chord cuts off two segments from a circle, the larger being termed *a major segment* and the smaller *a minor segment.* Their arcs are also called *major* and *minor* arcs.

   i) With radius r and θ the angle subtended at the centre of the circle, the area A is given by

$$A = \tfrac{1}{2}r^2\,(\theta - \sin\theta)$$

   ii) Where the length of chord c and the height of chord h are known, the area A will be given by

$$A = \tfrac{2}{3}\,\mathrm{ch} + \frac{h^3}{2c}\ \text{(nearly)}$$

   However, if the segment is flat, the second part of the expression can be ignored, giving $A = \tfrac{2}{3}\,\mathrm{ch}$.

f) **An ellipse** *is a curve traced out by a point which moves in such a manner that the sum of its distances from two fixed points, termed FOCI, is constant.* The longer axis is termed *the major axis,* the shorter one *the minor axis.* If these are denoted by M and m respectively, the area of an ellipse is given by

$$A = \pi \times \frac{M}{2} \times \frac{m}{2} = \frac{\pi M m}{4}$$

and the perimeter by $\left(\dfrac{M}{2} + \dfrac{m}{2}\right)$ or $\dfrac{(M + m)}{2}$

3.17.4 Areas of irregular-sided figures (see Fig. 3.2)

Such areas are generally found lying between the boundary of the area being surveyed and the chain line, i.e. *the offset area.* Section 3.3 looks at the division of this area in general terms but rather more specific methods of dealing with the calculation of this and similar areas of land must be considered.

a) **Counting squares** is a suitable method often adopted for very small areas with highly irregular boundaries. The procedure is

   i) prepare a *grid of squares* on a piece of tracing paper or film to predetermined dimensions (e.g. 20m, 10m, 5m grid, etc.) and to the same scale as the drawing.
   ii) superimpose the grid over the area on the drawing under consideration.
   iii) count the number of squares within the boundaries of the area being measured. Part squares may be 'balanced' with one another, counted as $\tfrac{1}{4}, \tfrac{1}{2}, \tfrac{3}{4}$ squares, or even discounted if extremely small.

b) **Give and take lines** may be drawn on a plan to replace the boundaries for calculation purposes. They should be drawn such that the areas excluded approximate (by eye) to the areas included. Once the boundaries have been 'averaged out', the figure becomes a straight-sided type, the area of which can be calculated as outlined in Section 3.17.1.

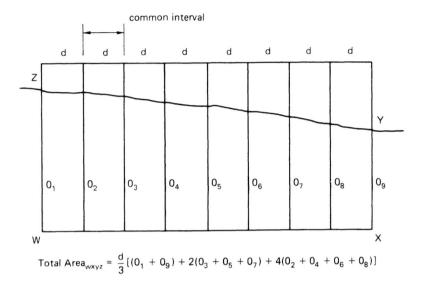

$$\text{Total Area}_{wxyz} = \frac{d}{3}[(O_1 + O_9) + 2(O_3 + O_5 + O_7) + 4(O_2 + O_4 + O_6 + O_8)]$$

Fig. 3.25 Simpson's Rule for areas

c) **Simpson's Rule** which enables the area of any curvilinear figure to be found to any degree of accuracy *provided the curve is a fair curve and the common interval is small.*
The rule is:

   i) by means of *ordinates, divide the figure* whose area is to be found into a *convenient number of parts of equal width* (see Fig. 3.25) which in turn will give an *odd number* of ordinates or heights.
   ii) *to the sum of the end ordinates, add twice the sum of the other odd ordinates and four times the sum of the even ordinates.*
   iii) *the total sum* thus obtained *multiplied by the common interval and divided by three* will give the required area.

    Thus, in Fig. 3.25, where d is the common interval between ordinates O_1, O_2, O_3 . . . . O_9, the area A of WXYZ will be given by

$$A = \frac{d}{3}[(O_1 + O_9) + 2(O_3 + O_5 + O_7) + 4(O_2 + O_4 + O_6 + O_8)]$$

In order to use Simpson's Rule, an *odd* number of ordinates is essential. If, for some reason, an even number of ordinates is given, ignore the first one, apply Simpson's Rule and then add the neglected area calculated by the *trapezoidal rule*. Simpson's Rule for Volumes is explained in Section 8.10.1.

d) **The Trapezoidal Rule** *enables the area of a curvilinear figure to be found approximately.*
The rule is:

   i) by means of *ordinates, divide the figure* whose area is to be found into *any convenient number of parts of equal width.*
   ii) *to half the sum of the first and last ordinates add the sum of the intermediate ordinates.*
   iii) *the total sum* thus obtained *multiplied by the common interval* will give the required area.

Thus in Fig. 3.25, the area A of WXYZ is given by

$$A = d[\frac{(O_1 + O_9)}{2} + O_2 + O_3 + O_4 + O_5 + O_6 + O_7 + O_8]$$

The Trapezoidal Rule for Volumes is explained in Section 8.10.1.

3.17.5 Special applications (see Figs 3.26 and 3.27)

a) Cross-section areas require to be calculated to facilitate the calculations of volumes (see Section 8.10) and all normal methods, triangles, counting squares, application of formulae and use of planimeter, may be employed. However, it is often more effective to apply appropriate formulae when dealing with long runs of uniform earthworks. Fig. 3.26 illustrates an embankment and a cutting and, in each case, h represents the height of the fill or cut, w the width, b the width of the formation and the sides slope at a gradient of 1 unit vertically in ℓ units horizontally.

  i) When the formation width b and the height/depth h are known, the width w will be given by

$$\frac{w}{2} - \frac{b}{2} = h\ell$$

$$\therefore \quad \frac{w}{2} = \frac{b}{2} + h\ell$$

      from which the full width of the works will be

$$w = b + 2h\ell$$

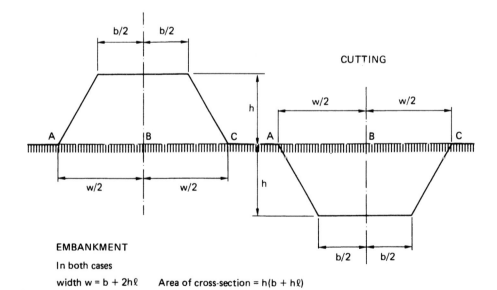

EMBANKMENT

In both cases

width $w = b + 2h\ell$ Area of cross-section $= h(b + h\ell)$

Fig. 3.26 Cross-section areas

ii) Having calculated the width, the area A of the cross-section is given by

$$A = \frac{(a + b)}{2}\ell \quad \text{the area of a trapezium}$$

$$\therefore \quad A = \frac{(w + b)}{2} \times h$$

$$= (b + \frac{2h\ell}{2} + b) \times h = \frac{(2b + 2h\ell)}{2} \times h$$

$$= h(b + h\ell)$$

iii) The formula may be applied when the *Reduced Level(RL)* at point B, the formation width and the slope of the sides are known *providing the section is level-across. A level-across section* is one in which the original ground surface is horizontal across the section through the works.

iv) When there is a cross-fall, or a section is part in cut and part in fill, or the original ground surface changes level and slope across the section, more complex formulae are available. However, the planimeter is a faster and simpler alternative unless computer programs are used.

b) **The use of co-ordinates** will enable the area within the traverse lines of a closed traverse survey to be calculated without the need to plot the survey (Fig. 3.27).

The rule is

The area within the lines of a closed traverse survey equals the algebraic sum of the products of the longitude of each line and the partial northing of that line.

The student should note that

i) *The longitude of a line* is the perpendicular distance from the meridian to the mid-point of the line.

ii) The area between any line and the meridian will be equal to *the longitude of the line multiplied by the partial northing of that line.* (The sign of the product will depend on the sign of the partial northing).

iii) If the area calculated has a -ve sign, ignore it, since it is the number value which is important.

iv) In practice, there is no need for any sketching or drawing to be carried out. Simply apply the rule from a knowledge of the partial co-ordinates of the lines of the closed traverse.

3.18 Summary

From the study of Chapters 2 and 3, it is possible to appreciate the whole concept of the chain-survey technique. It starts with a preliminary look at the site and ends with the completion of the drawing which can be presented to the client for his information and upon which other professionals can base their own work.

Not only must great care be exercised in the field but also equal attention must be paid to the interpretation and presentation of the gathered inform-ation in its drawn form. The competence of the surveyor must be matched by

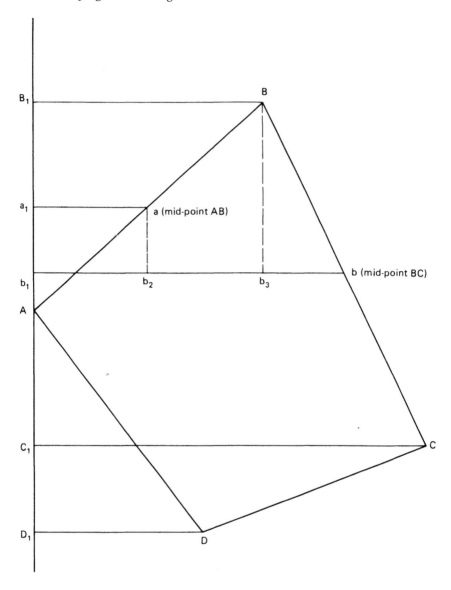

longitudinal line of AB = $b_1 b_2$ = $a_1 a$

Fig. 3.27 Areas using co-ordinates

the competence of the draughtsman. If they are one and the same person, that person will feel an immense satisfaction in carrying the process through from start to finish. When the technique is augmented by other techniques to include *height measurement* and *angle measurement*, the whole concept of surveying becomes one of great fascination.

Exercises on Chapter 3

1. *List* the nine major causes of error in chaining and, by means of brief notes, explain what effect each will have upon the survey and what precautions must be taken in order to eliminate these causes. When are errors in chaining discovered?
2. Describe how offsets are measured, with particular reference to the determination of 'how many' and 'where' such measurements are taken.
3. In a four-sided figure ABCD, formed by part of a boundary AB and part of a chain line DC, AD and BC are offsets at right angles to the chain line. AD measures 10.500 m, BC 12.100 m, and DC 10.000 m. What is the area of the trapezium ABCD? [113 m^2]
4. What is the difference between 20 m square and 20 square metres?
5. Explain fully, by means of sketches and notes, how a right angle may be set out from a point in a chain using a chain and/or tapes only.
6. Using a chain and tape, how would the following be set out in the field: (a) a 45° angle, (b) a 60° angle?
7. How would you carry out a parallel diversion?
8. Explain, with diagrams and notes, how to determine the distance between poles A and B on a chain line but on opposite banks of a wide river when the chain line is (a) at right angles to the river and (b) crosses it obliquely.
9. Explain the term 'chain angle' and show by diagrams how such a device is used.
10. Using the field notes in Fig. 3.28, plot the survey to a suitable scale.

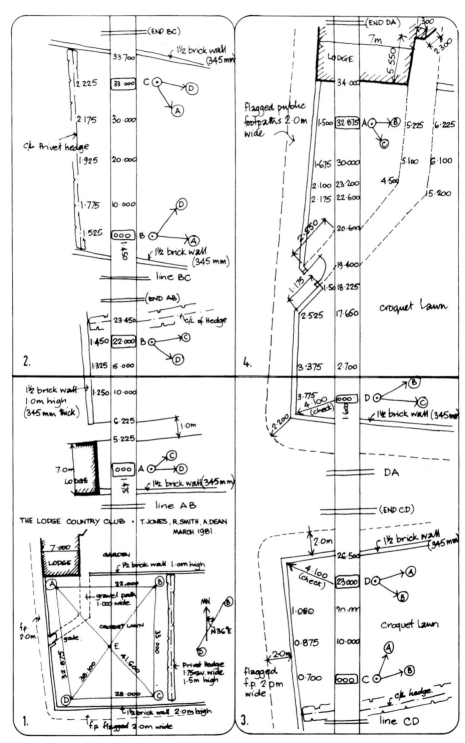

Fig. 3.28 Field-book notes relating to Exercise 10 and Fig. 3.19

4

Levelling – principles and equipment

When a small open site is to be surveyed and levelled, the proper sequence of operations is

i) carry out the physical horizontal measurement of the site using the chain-survey technique;
ii) carry out the *levelling operations*, including the calculation of the *reduced levels* of the points of ground to be 'levelled';
iii) plot the survey and note the reduced levels (and, if required, *contours* – see Section 5.12) either on the finished plan or on an overlay sheet.

This sequence is for what would be termed a full survey, and the student must appreciate that stages (i) and (ii) are not necessarily carried out in concert every time survey work is undertaken. Stage (i) may be carried out as a separate entity when height information is not necessary to the result. Similarly, where only height measurements are required, these may be taken with a minimum of horizontal measurement.

4.1 Reasons for levelling operations

Levelling is formally defined in Section 1.1. In the surveying of building sites, levelling operations are carried out primarily for the following reasons:

a) to determine the difference in height (altitude) between two or more points;
b) to determine the height (altitude) of the ground at a number of points along any desired lines so that *sections* may be drawn;
c) to set out level or horizontal surfaces such as floor slabs, foundation trenches, machine bases, etc.

It is often important to know whether ground is level or, if not, to what extent it varies in level. It can also be important to know the gradient of a sewer between two manholes or the difference in height of the top and bottom of an earth bank at various points along its length. Levelling techniques are therefore necessary in order that this type of question may be answered quickly and accurately.

Since the naked eye is very easily deceived, and cannot be relied upon for unassisted visual judgement, there is no adequate substitute for the use of instruments when measuring heights.

The basic equipment necessary is

i) an instrument capable of giving a *truly horizontal line of sight* – e.g. a level;
ii) a graduated staff for reading the *vertical height* from the ground or object to this horizontal line – e.g. a levelling staff;
iii) a special book in which to note the readings from which calculations can be made – e.g. a level book.

Where levelling operations are being carried out independently of a measured survey, the standard measuring equipment or chain, tape, etc. will be required in order to locate, on a map or plan, the points on the ground where readings have been taken.

From the results obtained by levelling operations, architects, engineers, quantity surveyors, and others can determine the suitability of a site for development; design buildings, roads, drainage, and other artefacts in relation to the contours of the ground; calculate areas of cut and fill; and so on, provided that the heights can be related to the horizontal plan and are produced accurately in the knowledge of why they are required.

4.2 Principles of ordinary levelling

The process of levelling may be *direct* or *indirect* (see Fig. 4.1).

a) **Direct levelling** is the method of taking a direct measurement, up or down, from one point to another. It is the method by which differences of height are measured, vertically, from a truly horizontal line of sight, established by various means as discussed in Section 4.3. The method is used by architects, engineers, surveyors, and builders for lower-order work.
b) **Indirect levelling** is the method of taking an indirect measurement by observing the angle of elevation or depression from one point to another. The tangent of this angle when multiplied by the horizontal distance apart of the points gives their difference in height (after corrections). This method is used in advanced levelling and higher-order work and is termed *trigonometrical levelling*.
c) **Barometric levelling** is a third method of finding the difference in height between two points, by means of simultaneous readings of barometric pressure at the two points.

Although the student must be aware of methods (b) and (c), which will be the subject of later studies, the present text will be confined to *direct levelling* by means of instruments generally using the spirit-level system of obtaining a horizontal line.

4.3 Methods of obtaining a horizontal line or plane

Ordinary levelling, as we have seen, is concerned with the measurement of vertical distance between a point on the ground and an arbitrary but *truly*

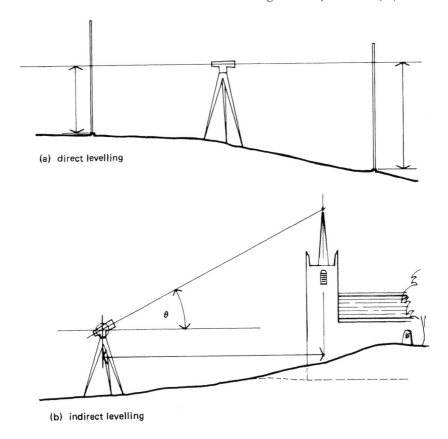

(a) direct levelling

(b) indirect levelling

Fig. 4.1 Methods of levelling

horizontal line or plane. But, for the student fully to understand what is meant by a *truly horizontal line,* two definitions are necessary at this juncture, to amplify those given in Section 1.2:

- i) a *level line* is a line that is of constant or uniform height relative to mean sea level and, because it follows the *mean surface* of the earth, it is a *curved* line;
- ii) a *horizontal line* through a point is *tangential* to the *level line* which passes through the same point and is *normal to* (i.e. at right angles to) *the direction of gravity* at that point.

These are illustrated by Fig. 4.2, and the student must appreciate the difference between a horizontal line and a level line. At the common point, the two lines will coincide; but, the greater the distance from the common point, the greater the discrepancy between the two lines. This is because one is truly horizontal and straight, while the other is curved and concentric with the mean surface of the earth (see Fig. 4.3(a)).

In ordinary levelling with sights less than 200 m, the difference is negligible and for all practical purposes can be ignored. Over long distances, however, calculations become necessary because the curvature of the earth will begin to affect the readings. Not only will the horizontal line and the level line diverge, but the line of sight could well leave the horizontal due to atmospheric

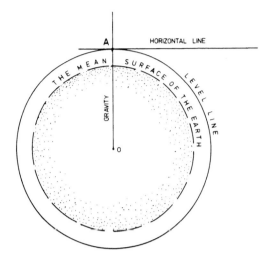

Fig. 4.2 The difference between a level line and a horizontal line

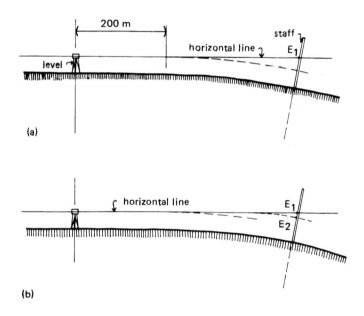

Fig. 4.3 Curvature and refraction

refraction. (*Refraction* is the term that is applied to the bending which occurs when rays of light pass through a medium having a non-uniform density. Near the earth's surface, the atmosphere is denser than it is at higher altitudes, with the result that rays of light are diverted *downwards*, so that in Fig. 4.3(b) the point actually observed is E_2 and not E_1.)

The prerequisite of ordinary levelling, therefore, is to obtain *a truly horizontal line at right angles to the direction of gravity*, and over the centuries many instruments and devices have been invented and used in order to obtain such a line.

The Romans used a device called a 'water level' (see Fig. 4.4). Basically this consists of a U-tube made of glass which is rigidly fixed to a support. When the U-tube is partly filled with water, the water surfaces in the two arms of the tube will have the same level and it is a simple matter to sight over the tops of these surfaces in order to obtain the required horizontal line of sight. Although this type of level persisted into the seventeenth and eighteenth centuries, its limitations are fairly obvious.

The author remembers watching an old 'drainer' using a different type of water level in the middle of a field. This was a box-shaped construction about

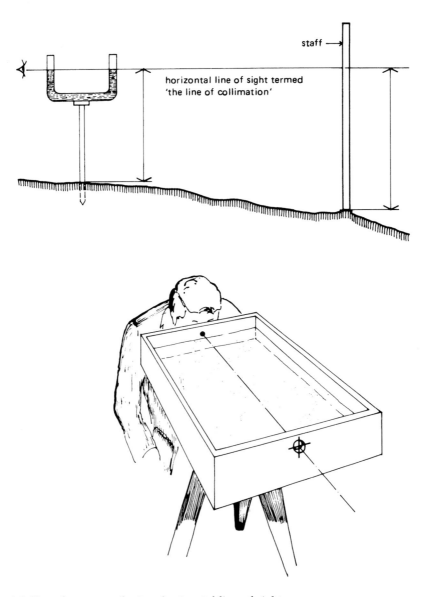

staff

horizontal line of sight termed
'the line of collimation'

Fig. 4.4 Use of water to obtain a horizontal line of sight

600 mm long and 150 mm deep which had been fixed to an ordinary tripod. A fine line had been scored around the inside of the open-topped box about 50 mm above the bottom, and the remaining features were a small hole bored at each end, one slightly larger than the other, just above the line. The smaller hole was the aperture, while the other had a crosswire at exactly the same height as the aperture. The drainer simply filled the box with water until the line marked round the inside was reached. The water was his levelling device and, amazingly, it worked for sights up to 50 m (see Fig. 4.4). When asked what he called his contraption, the old man solemnly touched his cap and said, "Tis a dumpy level, sir!'

The limitations and inaccuracies of a water level may not matter in terms of some aged retainer laying drains in a farmer's field, but for many other tasks greater accuracy is obviously required.

The force of gravity can be used in various simple mechanisms in order to define a horizontal line. A weighted pendulum, when freely suspended, defines the direction of gravity and, if a cross-piece is accurately fixed at right angles to the pendulum, the required horizontal line is obtained by sighting along the cross-piece (see Fig. 4.5(a)). As an alternative, a plumb-bob and line may be attached to the centre of a large inverted protractor. When the string coincides with the 90° mark, the straight edge of the protractor is horizontal (Fig. 4.5(b)).

While somewhat clumsy and inaccurate, the plumb-bob and line has always been used in this way, while a refinement of the pendulum principle is still the basis of some modern self-levelling instruments.

There is no doubt, however, that the simplest and most effective device for defining a horizontal line is a spirit-level tube. First developed in the mid seventeenth century, the early versions were in the form of a glass tube, bent

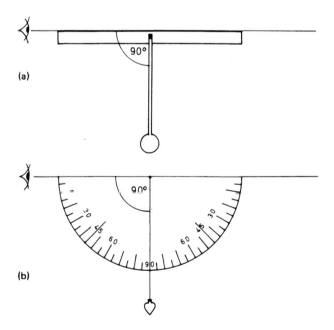

Fig. 4.5 Use of a pendulum and plumb-bob to obtain a horizontal line of sight

to a slight curve and partly filled with fluid. The resulting bubble of trapped air would always seek to find the highest part of the tube. When such a bubble is centred in the length of the tube, the *longitudinal axis of the tube* will be horizontal. This is because the bubble-tube is part of a large circle and the longitudinal axis is tangential to the mid point, or zero graduations on the tube, about which the bubble must be equispaced for it to be said to be *'in the centre of its run'* (see Fig. 4.6(a)). Any present-day builder's site level will illustrate the principle.

When a bubble-tube is attached to a straight-edge, levels may be transferred over short distances only, up to say 1 metre (Fig. 4.6(b)). The author's aged drainer friend, however, very often used a builder's level on top of a music stand and sighted along this to his staff (Fig. 4.6(c)), but this method is again limited by the distance that can be observed by the naked eye.

It was obvious that the limitations of the naked eye had to be overcome if the graduations on a levelling staff were to be read over long distances. It was found that this could be achieved reasonably easily by attaching the bubble-tube to a telescope which had a cross-wire within it to act as a sight for the purpose of aiming at the staff.

This arrangement has been the basis of the surveyor's level for over two centuries, and the markings on a graduated staff, when observed through the

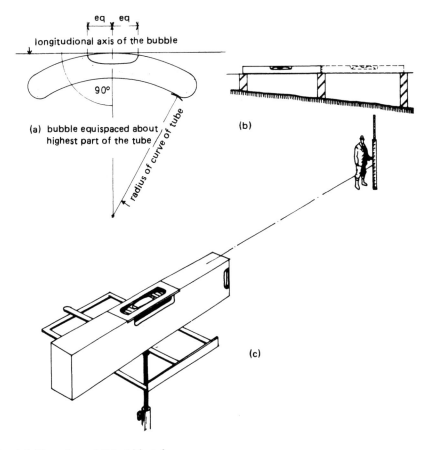

Fig. 4.6 Use of a spirit bubble-tube

telescope, will give the vertical distance from a point on the ground to the horizontal line of collimation – provided that·

a) the telescope has a cross-hair sight;
b) the spirit bubble-tube is attached to the telescope in such a way that the line of sight and the bubble-tube axis are parallel; and
c) the instrument has a stable support (e.g. a tripod) with provision for adjusting the instrument so that the bubble-tube axis becomes horizontal and hence the line of sight (the *line of collimation*) is also horizontal.

4.4 Principles of the level

The instrument used for ordinary levelling is called a *level* and, of the many types that have been invented or designed from time to time, comparatively few have survived except as museum pieces. It is sufficient, therefore, to look in detail only at the types in general use today.

Levelling is not a difficult operation but, for proper understanding of all the principles involved, there is no substitute for practice in the field. The surveyor must be thoroughly familiar with the instrument in all its parts so that, besides being able to set up and use it in the field, he can also test it and adjust it when necessary. In other words, a thorough knowledge of the principle and construction of the level is required.

4.4.1 The telescope

The first telescope was invented by a Dutchman, Jan Lippershey, in 1608, and the great mathematician Johannes Kepler suggested how the device could be developed for use in surveying instruments. This type of telescope is known as the *Keplerian* or *astronomical* type and consists of a tube of variable length which has an objective lens nearest to the object being observed and an eyepiece lens system at the viewing end. An *inverted image* of the object is formed by the objective lens and is in turn magnified by the eyepiece and presented to the observer in this inverted or upside-down form.

The fact that the world is seen upside-down and reversed poses no problems in practice, and one soon becomes accustomed to the phenomenon. Most modern instruments, however, are *terrestrial*, which is to say that, by the addition of a more complex eyepiece, an *erect image* is presented to the observer.

The early telescopes had a system of sliding tubes, and this *external-focusing* arrangement often caused imbalance in the instrument. The tubes also sagged, allowing dirt to enter easily. Modern instruments use an *internal-focussing* telescope, in which an extra lens has been introduced between the objective lens and the eyepiece. The telescope is focussed by moving this lens by means of a rack and pinion within the tube, which enables the length of the telescope to be kept constant and sealed against the entry of dirt and dust (see Fig. 4.7).

Modern lenses are 'bloomed' in order to reduce reflection and allow more light to pass through them and are of very high performance.

The most important considerations of any telescope are its magnification and resolving powers, i.e. the ability to show fine detail. In simple terms, the

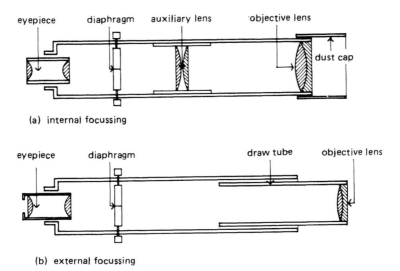

(a) internal focussing

(b) external focussing

Fig. 4.7 Types of telescopes

lens brings the object near to the observer and the magnifier (i.e. the eyepiece) enlarges it for inspection. Magnification can vary from × 10 to × 40 or more.

4.4.2 The line of collimation

The remaining part of the telescope requiring description is the *diaphragm*, which is also termed the *graticule*. An important feature of the Keplerian telescope is that an object positioned within the telescope at the focal plane of the eyepiece can be viewed clearly as a distant object is viewed and focussed. In earlier instruments the *diaphragm* was such an object and took the form of a metal ring with a strand of hair glued across its diameter. The hair was later replaced by a strand of spider's web. Obviously these could not be positioned with any great accuracy, so adjusting screws were necessary to position the cross-hair precisely. Added to this, the cross-hair was easily affected by a damp atmosphere and either sagged or snapped and, once broken, was not readily replaced. The modern instrument no longer uses the thread of the garden spider but has a diaphragm carrying a *reticule* (i.e. a network) of fine lines etched on glass.

Originally, one single horizontal line was used to mark the line of collimation. To this was added a vertical line (at right angles to the cross-hair) to ensure that the observation was made through the centre of the objective lens and also as a guide to the observer that the staff was in fact vertical. (Note: Signal 6 in Fig. 2.7 may be adopted by the surveyor to ensure that his assistant is holding the staff vertical. Signals 7, 8, 9, and 10 may also be used in order to move the staff position if observation is obstructed etc.)

Two further additions to the reticule were made in the form of short horizontal lines, equidistant about the line of collimation. These are termed *stadia lines* and are used for *distance measurement* called *tacheometry* (see Section 5.3). Various reticule patterns are shown in Fig. 4.8.

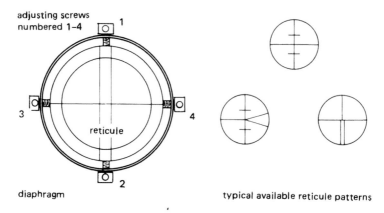

adjusting screws numbered 1–4

reticule

diaphragm

typical available reticule patterns

Fig. 4.8 Diaphragm and reticule

The separation required between the reticule and the eyepiece of the telescope depends on the eyesight of the observer. The lines on the reticule need to be brought into sharp focus (i.e. clear and black) before observations are made. This is done by gently screwing the eyepiece in or out and is known as *eliminating parallax*. Once set, the instrument is then satisfactory *for that observer* for the rest of the day. (See also Section 5.5.3.)

Finally, the beginner always finds it necessary to close one eye when sighting through a telescope. This can cause considerable strain over a full day's work, and the sooner the student can observe with both eyes open the better, since this is far less tiring and time-consuming.

4.4.3 The spirit bubble-tube

The bubble-tube is always fixed to the telescope barrel, either at one side or on top. It is impossible for this fixing to ensure that the bubble-tube axis will always be exactly parallel to the line of collimation, therefore it is usually bolted to the barrel at one end and fixed by means of an adjustable screw at the other.

The sensitivity of any bubble-tube depends on the curvature of the tube – the flatter the curve, the more sensitive the bubble, enabling it to be levelled more precisely. Sensitivity is described in terms of the amount of angular tilt of the longitudinal axis required to make the bubble travel a distance of 2 mm along the tube. The most sensitive will be about 10 seconds of arc per 2 mm, varying up to less sensitive types of 60 seconds of arc per 2 mm.

Various methods are employed for noting the bubble position. In the early instruments it was simply a case of walking round the instrument and observing, by eye, the bubble in relation to graduations marked on the tube. The most common later development, which is still used in modern instruments, was to place a mirror at 45° above the tube so that the observer could see the bubble position without moving from the eyepiece end.

However, since the bubble cannot be centred very accurately by the naked eye, a further development is the *coincidence-prism reading system*. A mirror is placed below the tube and deflects light through the tube to a series of prisms. These, in turn, project an image of half of each end of the bubble to an eyepiece viewer. When the two split ends of the bubble are coincident (i.e. appear to be in-line), the bubble-tube has been accurately levelled. This system is fitted to most modern instruments, especially those used for high-accuracy work, and can be read either in the bottom of the eyepiece or by means of a separate viewing window. If the system has a disadvantage it is that it is not always clear which way the level needs to be tilted to bring the images together. Different manufacturers have devised methods such as arrows, coloured fluids, etc. to overcome this problem.

Figure 4.9 illustrates various bubble-reading systems.

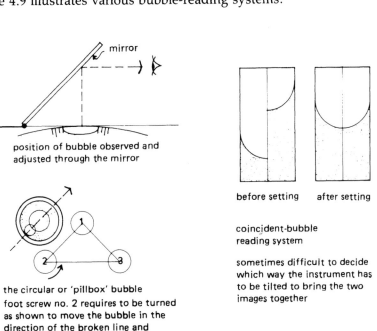

position of bubble observed and
adjusted through the mirror

the circular or 'pillbox' bubble
foot screw no. 2 requires to be turned
as shown to move the bubble in the
direction of the broken line and
hence to the centre of the circle.

before setting   after setting

coincident-bubble
reading system

sometimes difficult to decide
which way the instrument has
to be tilted to bring the two
images together

Fig. 4.9 Bubble-reading systems

4.5 Types of level

No matter how individual manufacturers interpreted the basic principles of the level, all tried to overcome the difficulties inherent in attempting to make the bubble-tube axis and the line of collimation parallel and horizontal at the moment the staff is observed.

Probably the oldest type of level is that known as the *Y-level* (or Wye level) which derives its name from the shape of the supports in which the telescope rests. The telescope, complete with its bubble-tube, can be removed from the collars at the top of the supports and replaced end for end, besides being rolled over within the collars (i.e. revolved about its horizontal axis). In the past, this level was used for precise work because a reading could be taken in four distinct positions and the mean of the four readings would give the correct observation irrespective of any inaccuracies in the instrument. Once the collars became worn, however, the instrument lost its efficiency. Although now a museum piece, the instrument was very important in its day, being favoured by opticians because it was capable of giving very accurate results, despite its somewhat loose construction, which was liable to throw it out of adjustment.

The three types of instrument used for ordinary levelling today are the *dumpy level*, the *tilting level*, and the *automatic level*, although this last does not use a spirit-level system.

4.5.1 The dumpy level (Fig. 4.10)

This instrument became the natural successor to the Y-level for ordinary levelling. Originally the name 'dumpy' was applied to the level invented by William Gravatt, which had a short and rather thick telescope which contained a large-aperture short-focus objective lens. The name 'dumpy' no longer has any bearing on the size or shape of the level, and most small levels on a building site are referred to as dumpy levels – more often than not incorrectly.

The dumpy-level consists essentially of the following parts: (a) a tripod, (b) a parallel-plate system, (c) a body piece, (d) a telescope, and (e) a bubble-tube.

a) **The tripod**   This consists of three wooden legs, either solid or framed, and usually telescopic for ease of carrying. Each leg is attached to the tripod head by means of a hinged joint. The tripod head itself may form the *lower part* of the *tribrach* (i.e. the base of the instrument) or there may be a screw fitting which will accept the whole instrument.

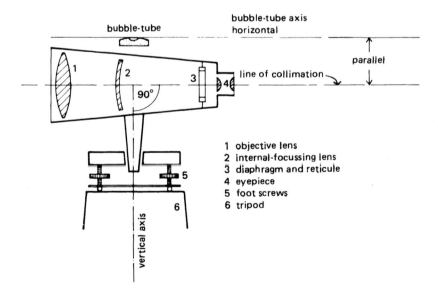

Fig. 4.10 A dumpy level

b) The parallel-plate system This consists of two circular plates, kept at a fixed distance apart by means of three or, in the case of a very old instruments, four screws known as foot screws. Each foot screw has its lower end widened and fashioned into a ball which fits into a small socket in the lower parallel plate, known as the *lower tribrach*. The upper parallel plate is simply a tribrach through which pass the foot screws. By turning these screws, the instrument is levelled (see Section 5.5.2).

c) The body piece and the vertical spindle In early versions of the instrument, the body piece consisted of a flat horizontal bar, to the ends of which were fixed two vertical pieces as supports for the telescope. The bar was fixed rigidly to a vertical spindle which fitted on to the upper plate, rotating within it to sweep out the horizontal plane.

Most modern dumpy levels have the telescope barrel, body piece, and vertical spindle cast in one piece and then machined to shape. This gives great strength and rigidity, and even lightness, since this method of manufacture dispenses with many small parts. The vertical spindle is still supported by the upper plate and able to rotate within it in order to sweep out the horizontal plane; while the vertical axis, and hence the telescope and bubble-tube, may be tilted in any direction by the appropriate rotation of the footscrews.

d) The telescope This has been discussed earlier. In the older instruments it would be an external-focussing type, but modern dumpy levels use the internal-focussing arrangement. In both cases, the telescope is rigidly fixed to the vertical axis, or spindle, by means of the bar or casting.

e) The bubble-tube This must be attached with *its axis at right angles to the vertical axis*. This will ensure that, when the vertical axis, through the spindle, is set truly vertical by manipulation of the foot screws, the bubble axis will be horizontal. Provided that the line of collimation *is parallel to* the bubble axis, rotation of the instrument about its vertical axis will result in the line of collimation sweeping out a horizontal plane.

The action of making the vertical axis truly vertical is termed *levelling-up* and will be looked at in detail in Section 5.5.2.

From the above, it is obvious that two conditions are critical:

   i) the bubble-tube axis must be at right angles to the vertical axis;
   ii) the line of collimation must be parallel to the bubble-tube axis.

If either condition is not fulfilled, accurate work is out of the question. The instrument should be checked regularly to ensure that it is in proper working order.

Besides having an internal telescope, the modern dumpy level is fitted with a bubble mirror and an engraved glass reticule, while some instruments will be found to be equipped with a horizontal circle of degrees for occasional angle measurement.

The student may come across the older instrument which has four foot screws, an external-focussing telescope, spider's-web cross-hairs with four adjusting screws, and no bubble mirror. Unless it has been in regular use, it will require renovation, but at least it may be quite valuable as an antique, especially if it contains a fair amount of brasswork!

4.5.2 The tilting level (Fig. 4.11)

Just as the dumpy supplanted the Y-level, so the tilting level became more popular than the dumpy. The essential parts of the instrument are (a) a tripod, (b) a support system, (c) a telescope and pivot, (d) a tilting screw and buffer spring, and (e) a bubble-tube.

a) The tripod　This is exactly as for the dumpy level, except that the head may have to receive a different support system to the normal screw.

b) The support system. The vertical-axis spindle can be supported in either of two ways:

　　i) by a horizontal plate with foot screws;
　　ii) by a ball-and-socket joint arrangement.

In either case, the top of the vertical axis carries a *stage* which supports the telescope. The vertical-axis spindle, stage, and. telescope can be rotated through 360° as one unit.

c) The telescope　This is of the internal-focussing type. It is *not rigidly fixed* to the stage and spindle but is attached to the stage by means of a pivot or hinge. This arrangement allows the telescope to be tilted at an angle to both the stage *and* the vertical axis. The telescope is therefore *not maintained at 90°* to the vertical axis as in the dumpy level.

d) The tilting screw and buffer spring　The tilting of the level about the pivot joint is carried out by means of a *tilting screw*. This passes through the stage by means of a threaded hole and bears against the underside of the telescope near the eyepiece end. Towards the objective end of the stage, a buffer spring is fixed which pushes up against the underside of the telescope. When the screw is wound vertically upwards, the objective end of the telescope will tilt downwards and compress the spring. When the screw is wound downwards

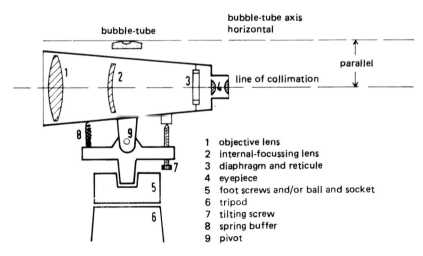

1 objective lens
2 internal-focussing lens
3 diaphragm and reticule
4 eyepiece
5 foot screws and/or ball and socket
6 tripod
7 tilting screw
8 spring buffer
9 pivot

Fig. 4.11 A tilting level

the reverse will apply, the spring being released and the objective end of the telescope being pushed upwards.

e) The bubble-tube As in the dumpy level, the line of collimation and the bubble axis must always be parallel.

In the field, the instrument is roughly levelled up by either the foot screws or the ball and socket, whichever is fitted. The ball-and-socket system has a small circular spirit-level, known as a *pillbox level* or *plate bubble*, fitted to the stage to facilitate the rough levelling-up. A level fitted with a ball and socket is often known as a *quick-set*. Rough levelling takes up very little time, but it is *essential that the bubble is centred by the tilting screw immediately before every observation*. The reading system is normally by means of a split bubble, through either the eyepiece or a separate viewer.

It is obvious, therefore, that the tilting level has only one critical condition, which is that *the line of collimation and the bubble axis must be parallel*.

The popularity of the tilting level is for two reasons:

i) the speed of setting-up and operation;
ii) the reduction in errors due to maladjustment, since there is only one *permanent adjustment* required as against two for the dumpy level (see Sections 4.6 and 5.20).

In most cases, levels for high-order work will be of the tilting type, although these instruments may be used for all classes of work. They may be equipped with a horizontal circle of degrees, and the tilting screw may be graduated in such a way that gradients can be set out. It is then termed a *gradienter screw*. In some instruments, the telescope can be rotated about its longitudinal axis, in which case it would be termed a *reversible* level and would be used for work requiring the mean of several readings in order to eliminate errors. The action of centring the bubble by means of the tilting screw at every sight has little or no effect upon the height of the line of collimation and only takes a second or so.

4.5.3 The automatic level (Fig. 4.12)

This third type of level which is in everyday use does not use the spirit-level but relies upon the action of a complex pendulum-and-prism device to define a horizontal line of collimation automatically without manipulation.

As in the dumpy level, the telescope is rigidly fixed to the spindle or vertical axis and may be supported on either foot screws or a ball and socket on a tripod. It does not require a tilting screw.

Once the instrument has been levelled up to within ± 10 minutes or better, rotation of the telescope will always sweep out a horizontal plane. Although manufacturers' systems vary, the principle is that a horizontal ray of light entering the centre of the objective glass is passed through a system of fixed and suspended prisms and is directed by them to the centre of the cross-hairs on the diaphragm, being then observed through the usual eyepiece. The telescope is usually more powerful than either the dumpy or the tilting types, with better resolution and accuracy of levelling. The fieldwork is very fast and, because of the prism system, an erect image is given.

This type of level has only one permanent adjustment, namely that *the line of collimation is horizontal when the circular bubble is centred*.

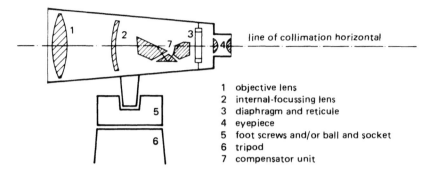

1 objective lens
2 internal-focussing lens
3 diaphragm and reticule
4 eyepiece
5 foot screws and/or ball and socket
6 tripod
7 compensator unit

Fig. 4.12 An automatic level

The instrument is quite robust in construction and is also *self-damping*, which is to say that the free movement of the suspended prism system is quickly and automatically arrested following movement of the instrument, without any mechanical control being applied by the surveyor (see Section 5.5.4(c)).

4.5.4 General points

All telescopes rotate through 360° around the vertical axis in order to sweep out the horizontal plane. If this movement was unrestricted at all times, observing the staff would be a difficult and rather inaccurate business. Instruments are therefore fitted with clamps and screws which allow the rotation of the telescope to be arrested and then finely tuned.

Once the telescope has been aimed at and has found the staff, the *horizontal clamp* or *training clamp* is locked on. This will keep the telescope trained on the staff until such time as the clamp is released.

When the observation is made, fine adjustment to the aim may be carried out by means of a *tangent screw*. This will move the telescope in the horizontal plane, very slowly and within certain small limits. In some modern levels, the normal horizontal clamp is replaced by a *friction brake*. The telescope can thus be turned in any required direction and is held in this position by friction but can still be brought into proper alignment by a tangent screw.

Where levels are fitted with features such as a horizontal circle of degrees, differing reading systems may be employed by different manufacturers. It is therefore vital that the instruction booklet, which accompanies the instrument when it is first bought, is not only read before the level is used but is also kept safely in the instrument case for frequent consultation. These booklets give the operator full information about the construction of the level, its use in the field, how to test and check for maladjustment and how to adjust, and many other tips on surveying and are worth their weight in gold.

4.6 Instrument adjustments

The adjustments of the level are divided into two classes:

a) **temporary adjustments**, which are the operations required to set up the instrument in the field ready to start the day's work;
b) **permanent adjustments**, which are necessary from time to time to keep the instrument in working order. These adjustments should be carried out only when it is absolutely necessary, and in some extreme cases only by the instrument maker.

There are two permanent adjustments to be considered, namely

i) adjustment of the sensitive level (bubble-tube);
ii) adjustment of the line of collimation.

In the case of the dumpy level, these are *two separate operations*; while in the case of the tilting level the adjustments are carried out as *one operation*.

Detailed consideration of the above classes of adjustment will be covered in Chapter 5, since both require some knowledge of fieldwork before they can be properly carried out. To keep the learning process in some proper sequence, therefore, it is sufficient at this stage for the student to know that the level can become inaccurate and that adjustments can be made to restore the level to correct working order.

4.7 Levelling staves

The summary of BS 4484:part 1, 'Measuring instruments for constructional works', defines a levelling staff (plural '*staves*') as 'a staff consisting of wood or suitable metallic or synthetic material, graduated for vertical measurement, and read with an optical instrument. It may also be used for tacheometric measurement.'

As with the level, various types of staff have been used over the years, and Fig. 4.13 shows some of the arrangements that are currently in use for the standard 3 m and 4 m lengths.

The old heavy wooden Sopwith staff has been largely replaced by lighter easier-to-handle staves, which are equally satisfactory if handled with care. During the change-over from imperial measurement to metric measurement, it was not uncommon for perfectly sound imperial staves to be converted. An adhesive strip of metric graduations was applied to the face of the staff, and the student may well come across an old Sopwith staff reading in metric units.

a) *Wooden staves*   Generally these are made from well seasoned mahogany. While not subject to temperature change or variation, they do suffer from the influence of water (i.e. sections can swell and telescopic arrangements will not slide).
b) *Metal staves*   These are made from aluminium alloy and, while unaffected by water, they can suffer variation due to temperature changes. Perhaps the wooden staff has been more popular because it is not as cold to handle in winter as its metal counterpart.

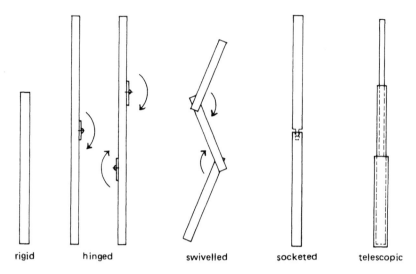

Fig. 4.13 Staff arrangements

A telescopic staff usually consists of two or three box sections which slide inside one another. When extended, the sections are held in place by a spring catch, and this must be heard to click into place as an indication that the section is fully extended and locked securely, and that correct readings will be attainable.

Socketed sections must be locked together with care for the same reasons.

When carrying the staff from position to position, it should be carried either almost vertically, using both hands and a shoulder for support, or, if fully extended, over one shoulder – provided that the staff is on its edge *and not on its face or back*, which would induce droop and strain the locking devices.

Graduations must be protected from wear and should be cleaned regularly to ensure that they can be clearly read at all times when viewed through the instrument. The patterns available are many and varied (one manufacturer has sixteen patterns in his catalogue) and it will suffice, therefore, to consider only the most common patterns used in ordinary levelling (see Fig. 4.14). The metric 'E' pattern (to BS 4484) consists of metres, decimetres, and centimetres, and millimetres by estimation. The staff may be observed as an erect image or as an inverted image, and the pattern may be printed accordingly. The student should therefore familiarise him/herself with the staff and the level before starting field operations.

Finally, the quality and the condition of the staff are very important. They should match the level with which the staff is to be used. There is not a lot of sense in using a £500 level with a £5 part-broken staff!

4.7.1 Accessories

There are various items which may be used with a staff to improve results or to ease the labour.

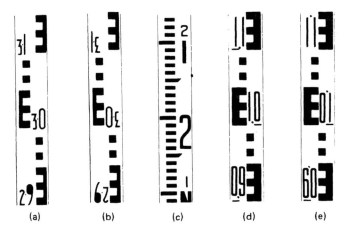

Fig. 4.14 Levelling-staff face patterns. The metric 'E' pattern is shown 'erect' at (a) and (d) and 'inverted' at (b) and (e). The 'inverted' form will read 'erect' through normal instrument telescopes, and vice versa. The simple pattern at (c) has m, dm, cm, and $\frac{1}{2}$ cm (5mm) divisions. In all cases, the student should note how the numbers relate to the division marks for reading purposes.

a) *Staff bubble*   This is a circular spirit-level which is attached to either the side or the back of the staff. This will enable the staffman to improve the verticality of the staff while observations are being made.
b) *Staff holder*   This is a U-shaped clamp with two handles. Usually made of timber, it can be clamped to the staff to make it easier to hold vertical in windy conditions. Some staves may have handles fitted as standard, while stay-rods are also available.
c) *Change plate*   This is a triangular steel plate, with its centre raised and its three corners turned down. There is also a length of chain attached for carrying purposes. The device is used where a change point (see Section 5.1.9) occurs on soft ground. It is stamped firmly into the ground to form a stable base which will ensure that the foot of the staff remains at the same height above datum for both backsight and foresight readings.
d) *Bench-mark staff bracket*   This is used where the Ordnance Survey bench-mark (see Section 5.1.3) is flush with the wall.

4.8 General

The overriding significant point which comes out of the above text is that the student must know and understand the instruments before he can put them to effective practical use in the field. It is therefore necessary to read the instruction booklets supplied with new levels and to acquire manufacturers' catalogues to appreciate the full range of instruments and equipment available to the modern surveyor.

Exercises on Chapter 4
1. What is the proper sequence of operations when a small site is to be surveyed and levelled?
2. By means of sketches and notes, explain the difference between direct and indirect levelling.
3. Explain the terms (a) level line, (b) line of collimation.
4. Describe the temporary adjustments of a dumpy level.
5. Explain the following terms: (a) terrestrial, (b) parallax, (c) permanent adjustments.
6. Explain the differences between a tilting level and an automatic level.
7. What are the critical conditions of the dumpy, tilting, and automatic levels so that they may be used for accurate work?
8. What are the functions of a tangent screw and clamp?
9. By means of diagrams, show the different staff arrangements available and explain what is meant by the metric 'E' pattern.
10. List, with sketches and brief notes, four accessories which may be used in conjunction with a levelling staff.

5

Practical levelling and contouring

With an understanding of the equipment necessary to the carrying out of levelling operations, further definitions must now be given as a basis for the proper understanding of the fieldwork involved.

5.1 Definitions

5.1.1 Level surfaces

A level surface is a surface that is everywhere at right angles to the direction of gravity of the earth. For all practical purposes it may be considered to be a *spherical surface* with its centre at the centre of the earth. Perhaps the best example is the surface taken up by any large sheet of still water, such as a lake. It follows from the definition of a *level line* (see Sections 1.2 and 4.3), that a level surface is composed of adjacent level lines and, conversely, that a level line is any line lying in a level surface.

5.1.2 Datums

A datum may be a *surface* or a *line* to which observed heights are related.

When two readings are taken at each end of a uniform slope, their heights can only be given relative to each other. By simply subtracting one reading from the other, the vertical difference in height between the two points can be stated. It can also be stated that point A is higher or lower than point B. If the horizontal distance between the two points is measured, the gradient of the slope between the two points can be calculated, as the vertical height between the two points is known. But that is all: there is nothing by which the two points can be related to the remainder of the site (see Fig. 5.1).

In order to relate a *series of heights* to each other, they must be given relative to a common point or plane known as a *datum*, which may be of two types: the ordnance datum or an assumed datum.

a) **The ordnance datum** (O.D.) is the datum to which all heights shown on *Ordnance Survey* (O.S.) maps are referred. This datum line is the *mean sea*

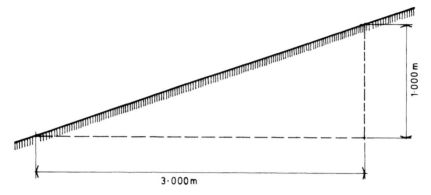

Fig. 5.1 Gradient 1 in 3

level at Newlyn in Cornwall, which was calculated from the results of many hourly observations taken principally between the years 1915 and 1921.

This O.D. replaced the *Liverpool datum*, which had been established in 1844 after only a fortnight's observation and was thought to be unreliable. The student should therefore be wary of using very old O.S. sheets, certainly pre-1915, which would be related to this now defunct O.S. datum, besides being in imperial units. (Note: Conversion from the Liverpool datum to the Newlyn datum is not easy, since there is a variance throughout the country and a constant factor cannot be used.)

Wherever it is convenient, the O.S. datum or *zero plane* is used when calculating site levels (see Fig. 5.3), although any other plane, whether real or imaginary, can be taken as the basis of operations. The height (altitude) of a point is its *distance vertically above the given level surface*; for example, '100 m O.D.' means that the point in question is 100 metres above ordnance datum and would be shown on an O.S. map as 'B.M. 100.00'.

b) **Assumed datum** Such a datum is used where it is inconvenient or impossible to relate the work in hand to the ordnance datum.

A point should be selected which is below the lowest level of the site and given the Reduced Level (R.L.) of 100.000 m. The reasons why 0.000 m is *not* used are

i) there can only be *one* 0.000 m datum and that is at Newlyn;

ii) if, for any reason, additional levels have to be taken which are just below the assumed datum, the R.L.'s of such points will be positive quantities and not negative (minus) as they would be if 0.000 m were used;

iii) because any R.L.'s below assumed datum are positive, there will be less likelihood of error in booking or calculation or mixing them up with *inverted levels* (see Section 5.16).

5.1.3 Bench-marks

A bench-mark is a fixed point of *known height* above the O.D. from which the height above O.D. of any other point may be determined. The various types of bench-mark, usually abbreviated to B.M., are illustrated in Fig. 5.2, and the student should note the position of the foot of the staff when readings are

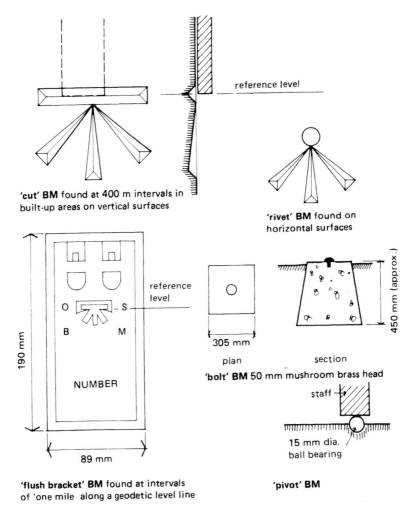

'cut' **BM** found at 400 m intervals in built-up areas on vertical surfaces

'rivet' **BM** found on horizontal surfaces

'flush bracket' **BM** found at intervals of 'one mile' along a geodetic level line

'bolt' **BM** 50 mm mushroom brass head

'pivot' **BM**

Fig. 5.2 Types of bench-mark

being observed on such a mark. Once the instrument has been set up, the first observation would be on a B.M., either an O.S.B.M. or the *assumed datum point* to which the levels are to be related.

5.1.4 Temporary benchmarks

A temporary bench-mark (T.B.M.) is a bench-mark set up by the surveyor for his own use for a particular task. The height of the T.B.M. may be established from an O.S.B.M. and will allow levels to be related to the O.S. datum and referred back to the T.B.M. without the often tedious checking back to the O.S.B.M. every time a series of levels is completed.

T.B.M.'s should be stable, semi-permanent marks, such as a wooden peg set in concrete, or some permanent feature of an existing building, e.g. 'top of plinth', 'front doorstep', and so on.

Note: The following terms used in levelling operations are illustrated in Figs 5.3 and 5.4 and must be thoroughly learned, since they will be constantly used throughout levelling operations and fieldwork.

5.1.5 Reduced levels

A reduced level is the height of a point or object stated with reference to the selected datum for the work in hand. It is abbreviated to R.L. Provided that the starting point of the operations is a known or assumed R.L., then the R.L.'s of the various points of the site can be calculated from this R.L. and the staff readings taken at the various points (see Section 5.2).

5.1.6 Backsights

A backsight is the *first sight, or reading*, taken after the instrument (the level) has been set up. This sight is taken to a point whose height is known, has been assumed, or can be calculated. It is abbreviated to B.S. and is taken at the start of the work and at a change point (see Section 5.1.9).

5.1.7 Foresights

A foresight is the *last sight, or reading*, taken during levelling operations *before* the instrument is moved. It is abbreviated to F.S. and is taken at a change point and at the end of operations.

5.1.8 Intermediate sights

An intermediate sight is any sight, or reading, taken *between a B.S. and an F.S.* It is abbreviated to I.S. and is sometimes termed an *intersight*.

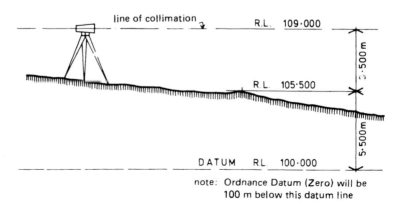

Fig. 5.3 Reduced level

5.1.9 Change points

A change point is an arbitrary point which enables the levelling to continue from a new instrument position. It is often also termed a *turning point* and is abbreviated to C.P. or T.P.

From the 'first' instrument position, an F.S. is taken, the instrument is then moved to the 'second' or 'new' position, and, following the setting up of the instrument, a B.S. is taken. *The foot of the staff must not be moved by the staffman from the time that the F.S. is taken to the time that the B.S. has been observed, booked, and checked.*

Both readings must be taken on exactly the same point of ground, and the only staff movement permitted is to carefully turn it so that the face can be read from the new instrument position. The student must appreciate that it is the instrument which moves, *not the surface of the earth*, and *a new horizontal line of sight is created by changing the position of the instrument*. It is absolutely vital that the two lines of collimation can be related accurately if the levelling is to be continued and be of use.

Closing errors at the end of the operations are generally the result of slipshod work at change points, and the amount of care required at this stage of the levelling cannot be too strongly emphasised.

In order to prevent any movement, the foot of the staff *must* be placed on solid ground or on some other hard surface, which may be off the line of levels being taken, provided its position is known (for later checking purposes). Where the ground is soft and a firm point is not readily available, the *change plate* as described in Section 4.7.1(c) should be used.

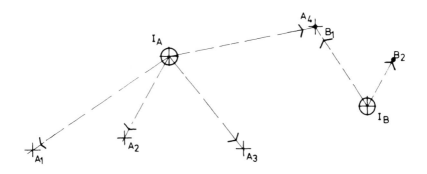

Fig. 5.4 Levelling terminology

With the instrument at station I_A:
A_1 is a *backsight* (the first reading),
A_2 and A_3 are *intermediate sights*,
A_4 is a *foresight* (the last reading taken from I_A before the instrument is moved to the next station I_B),
A_4/B_1 is a *change point*.

With the instrument moved and set up at I_B:
B_1 is a *backsight* (the first reading from this station),
B_2 is an *intermediate sight*.

5.1.10 The line of collimation

The line of collimation has already been described as the truly horizontal line of sight which passes through the optical centre of the telescope of the level. The height of this line above the datum is termed the *height of instrument* or *height of collimation*, while the horizontal plane swept out by this line as the telescope is revolved about its vertical axis is known as the *collimation plane* or *plane of collimation*.

5.2 Ordinary levelling

Most students find it hard initially to visualise just what constitutes the operation of levelling between two or more points. In practice, the principle is quite straightforward and certainly nothing to be frightened of, provided the operations are carried out in their proper sequence and with care.

The following examples should make the principle clear. For reasons of simplicity, the physical operations of setting up the instrument have been ignored here – they are discussed in Section 5.5.

Example 1   It is required to find the difference in height of two points A_1 and A_2, illustrated in Fig. 5.5, which are within working range of the level.

A convenient spot on firm ground is chosen which is roughly equidistant between the points (although not necessarily on-line between them) and

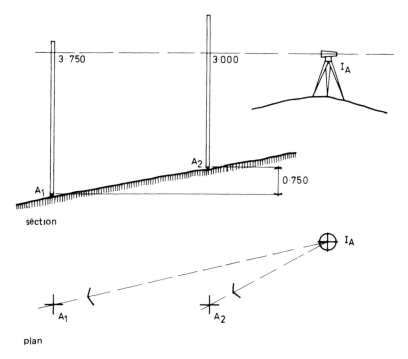

Fig. 5.5 Difference in height between two points

which allows an unobstructed view of the staff when it is held on each point in turn. The instrument is set up on this spot with the line of collimation truly horizontal and high enough for the staff to be read. An assistant places the staff on point A_1, holding it so that it is *vertical*. The surveyor directs the telescope of the level onto the staff and reads the figure where the staff is intersected by the horizontal line of sight. This figure is noted in a special book. The operation is then repeated with the staff held vertically on point A_2.

The readings, 3.750 m at A_1 and 3.000 m at A_2, are the respective vertical distances that each point is below the line of collimation, as shown in Fig. 5.5. By the simple subtraction of these figures, point A_1 is found to be 0.750 m below point A_2.

In levelling operations, however, the typical problem facing the surveyor is that the height of one point above datum is known and it is required to find the R.L.'s of other points above this datum.

Example 2 In Fig. 5.6, point A_1 is known to be 250.000 m above O.D. and the R.L.'s of points A_2, A_3, A_4, and A_5 are required to be found.

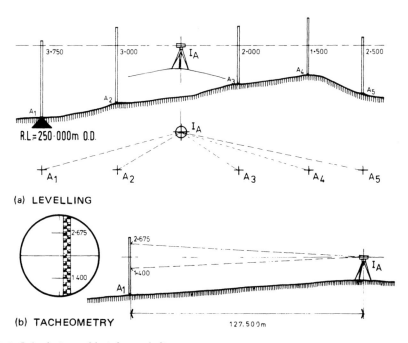

Fig. 5.6 Calculation of height and distance

Again the level is set up on a suitable spot from which the staff can be read as it is placed on each point in turn. The observations are made and the readings are booked in exactly the same way as the two points in example 1. The vertical distances measured from the ground to the line of collimation are now 3.750 m at A_1, 3.000 m at A_2, 2.000 m at A_3, 1.500 m at A_4, and 2.500 m at A_5. Since the R.L. at A_1 is known, the required R.L.'s can be calculated by either of the following two ways.

5.2.1 Method (a)– height of collimation

The first requirement is to establish the R.L. of the line of collimation. This is done by *adding* the distance measured from the line of collimation to the ground at point A_1 to the known R.L. of point A_1.

e.g. $3.750 \text{ m} + 250.000 \text{ m} = 253.750 \text{ m}$

This R.L. is also known as the height of instrument.

Once an instrument has been set up, unless it is kicked or similarly abused, the *height of instrument* will remain the same for each observation made until the instrument is moved to a new position. It follows then, that the R.L. of the line of collimation in the example will also be 253.750 m for each of the readings taken on points A_2, A_3, A_4, and A_5. *By simply subtracting each vertical distance from this figure, the R.L. of each point is given automatically;*

i.e. $253.750 \text{ m} - 3.000 \text{ m} = 250.750 \text{ m}$ R.L. at A_2

$253.750 \text{ m} - 2.000 \text{ m} = 251.750 \text{ m}$ R.L. at A_3

$253.750 \text{ m} - 1.500 \text{ m} = 252.250 \text{ m}$ R.L. at A_4

$253.750 \text{ m} - 2.500 \text{ m} = 251.250 \text{ m}$ R.L. at A_5

This is a simple, quick, and accurate way of calculating R.L.'s (which will be further discussed in later text) and is normally used for low-order work.

5.2.2 Method (b)– rise and fall

The difference in height between any two points is referred to as either as *rise* or a *fall* relative to one of the points. In the example shown in Fig. 5.6(a), point A_1 is nearer to the datum than point A_2, therefore the ground is rising from point A_1 up to point A_2. It is obvious, then, that point A_2 is at a greater height above datum than point A_1. The difference in height between the two points must therefore be *added* to the R.L. of point A_1 in order to determine the R.L. of point A_2.

In this method, the actual height of the line of collimation has no real significance other than being the line to which vertical distances are conveniently measured from various points on the ground, and the method relies upon the difference in height between successive points. *Each point is considered in relation to the point immediately preceding it and whose R.L. is either known or has just been calculated.*

The difference in height is obtained by subtracting the staff readings taken on the two points;

e.g. $A_1 - A_2 = 3.750 \text{ m} - 3.000 \text{ m} = 0.750 \text{ m}$ as in example 1

Since point A_1 now has a known R.L. of 250.000 m, by adding the difference 0.750 m to this figure the R.L. of point A_2 is found to be 250.750 m.

It is important to remember that the reading taken on the point whose R.L. is being calculated *is always subtracted from* the reading whose R.L. is known, or has just been calculated, in order to determine the difference in height between the two points. The reason for this is that, if the result of the calculation which gives the difference in height between any two points is *positive* then the difference is a *rise*; if it is *negative* then the difference is a *fall*. Since the difference in height is always *added* to the preceding known R.L. to determine the required R.L., the mathematical sign will always guarantee a correct result, provided the arithmetic is carried out correctly.

In the example shown in Fig. 5.6(a), the correct sequence of calculation is therefore

R.L. at A_1 is known

R.L. at $A_2 = (A_1 - A_2) +$ R.L. at A_1

R.L. at $A_3 = (A_2 - A_3) +$ R.L. at A_2

R.L. at $A_4 = (A_3 - A_4) +$ R.L. at A_3

R.L. at $A_5 = (A_4 - A_5) +$ R.L. at A_4

Let us now put the method to the test by using the readings from Fig. 5.6(a) and comparing the results with those given by method (a).

R.L. at A_1 = 250.000 m (known)

R.L. at $A_2 = (A_1 - A_2) +$ R.L. at A_1

$= (3.750$ m $- 3.000$ m$) + 250.000$ m

$= 0.750$ m $+ 250.000$ m

$= 250.750$ m (check)

R.L. at $A_3 = (A_2 - A_3) +$ R.L. at A_2

$= (3.000$ m $- 2.000$ m$) + 250.750$ m

$= 1.000$ m $+ 250.750$ m

$= 251.750$ m (check)

R.L. at $A_4 = (A_3 - A_4) +$ R.L. at A_3

$= (2.000$ m $- 1.500$ m$) + 251.750$ m

$= 0.500$ m $+ 251.750$ m

$= 252.250$ m (check)

R.L. at $A_5 = (A_4 - A_5) +$ R.L. at A_4

$= (1.500$ m $- 2.500$ m$) + 252.250$ m

$= - 1.000$ m $+ 252.250$ m

$= 251.250$ m (check)

The student should note particularly how the mathematical sign has worked in the calculation of the R.L. of point A_5. The *negative* sign indicates to the surveyor

a) that the ground has fallen from point A_4 to point A_5;
b) that point A_5 is therefore nearer to the datum than point A_4;
c) that point A_5, as a consequence, has a smaller R.L. than point A_4.

The results of the foregoing calculations bear out the above, and the student should remember that *when a negative difference is added to the preceding R.L., it has the effect of reducing that R.L.*, a fact which can readily be seen from the calculation of the R.L. of point A_5.

The significance of the two methods will be considered further when the subject of booking observations is discussed (section 5.6), but the student can see that *the same results were achieved by both methods*.

5.3 Tacheometry

A further use of the level and staff is to *measure distance* as opposed to height.

The technique used for this is called *tacheometry* and would not be possible without the *stadia lines* engraved on the glass reticule of the modern instrument (see Section 4.4.2). The simple *measurement of horizontal distance* needs to be considered briefly, since the technique is referred to elsewhere in the text.

Using the level, the surveyor will note the figures where the stadia lines cut the staff. The difference between these readings is termed the *staff intercept* (s) and, when this is multiplied by 100, the distance from the centre of the instrument to the face of the staff is given. (Although the constant of the modern instrument is normally 100, this should be checked in the instruction booklet – it may just be 50.) The distance between the stadia hairs on the reticule is fixed so that the further away from the instrument the staff is held, the greater is the length of intercept which is read and noted down. When taking readings, the staff may be held either *vertically or normal to the line of sight of the instrument*, although this latter is seldom used in practice.

The technique is best carried out using a tilting level. The tilting screw allows the *lower stadia line* to be brought to a whole decimetre (0.1 m) graduation of the staff, so that only the *upper stadia line* reading is estimated. This reduces reading error.

Example In Fig. 5.6(a), it is required to determine the horizontal distance from the level to point A_3.

The tilting screw is used to move the *lower* stadia line to read the whole decimetre graduation 1.400. The *upper* stadia line then reads 2.675 (see Fig. 5.6(b)).

Upper stadia line = 2.675

Lower stadia line = 1.400

∴   Staff intercept   = 1.275 m

Hence horizontal distance = 1.275 m × 100 = 127.500 m

The student must, however, ensure that if a distance is measured during the course of normal levelling operations *the longitudinal bubble axis needs to be reset parallel to the line of collimation before the levelling is continued* (refer to Section 4.5.2(e)).

Instruments used for measuring distance are often referred to as *Tacheometers* and in addition to using a tilting level, horizontal distances may be indirectly measured by *Theodolite* (see Chapter 6) and *electro-magnetic distance measurement (EDM) techniques* (see Chapter 10). However, when the ends of the 'line' being measured are on different levels, the technique used is a combination of tacheometry and trigonometrical levelling (see Section 1.4) no matter which type of instrument is being used.

5.4 Fieldwork procedure

No matter what the specific purpose of the levelling operations, a certain basic pattern is followed for ordinary levelling.

i) Obtain an O.S. map of the area (usually 1:1250 is adequate) and note all the O.S.B.M.'s near to the site. Ascertain whether or not any T.B.M.'s have been set up on the site which may be of use.

ii) Carry out a thorough reconnaissance of the site. This may be done before the chain survey if the levelling operations form part of a full survey. From this preliminary investigation, the methods to be employed will be determined.

iii) Select and prepare the equipment appropriate to the task. Set up the instrument at the first position and make ready for observing (see Section 5.5).

iv) Starting from the B.M. (O.S. or T.B.M.), observe the staff at all the points required. Book each reading in the *level book*, along with any remarks and notes necessary to the eventual plotting process.

v) If all the required points cannot be observed from the first instrument position, select a change point, move the instrument to a second position, set up, and continue observing and booking as in stage (iv).

vi) Repeat stage (v) as often as necessary to cover all the required points.

vii) Finally, either level back on to the B.M. on which the first reading was made, or level on to another B.M. of known height. If this is not done, there is no way that the accuracy of the levelling can be checked or gross errors be detected.

viii) In the office, the results are determined. The R.L.'s are calculated, adjustments are made if necessary, and the levels (and contours if required) are plotted on a master drawing to a suitable scale. The drawing in question may be the measured survey or an overlay.

5.5 Temporary adjustments

The setting up of the instrument – that is, firmly supporting it on stable ground with its line of collimation horizontal and ready for observing – is termed *carrying out the temporary adjustments of the level*. Although the actual operations may vary according to the type of level being used, the basic procedure is as follows.

5.5.1 The tripod

This must be firmly placed on stable ground. Most present-day tripods will be of the telescopic type and their legs may be adjusted to suit the ground and the person using the level. One leg point is placed on the ground and the other two legs are pulled towards the user and opened out. On flat ground, the ends of the legs should be equidistant from each other, while on sloping ground one leg will point uphill with the other two pointing downhill at equal intervals. The top of the tripod should be in the horizontal plane, to ease the operation of obtaining a horizontal line. This will mean that on sloping ground the leg pointing uphill will be shorter than the other two (see Fig. 5.7). Using the lugs fitted to the legs, push the feet of the tripod firmly into the ground. An even and steady pressure should be applied. When the feet are in place, check that all clamps and bolts are secure so that there is no danger of the tripod sinking or moving once levelling has begun.

The instrument is then taken carefully out of its box and attached to the head of the tripod. *Remember to check before leaving the office that the level does in fact fit the tripod.* Modern equipment is fitted with some form of captive bolt, screwed up into the underside of the instrument, although in older instruments the top of the tripod may be threaded and the level screwed on to this thread. Nothing is more frustrating when setting up than to find that the level and tripod will not 'marry'.

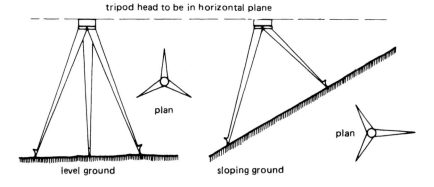

Fig. 5.7 Setting the tripod

5.5.2 Levelling up the instrument

a) **The dumpy level** Since it is unlikely that the student will use a four-screw level, only the levelling-up of the three-foot-screw type will be considered.

The telescope is placed parallel to two of the foot screws in plan as shown in Fig. 5.8(a) and the bubble is centred using these two screws. Each screw is held between forefinger and thumb and *both screws are turned simultaneously* at the same speed but in *opposite directions* – in other words, 'thumbs together' or 'thumbs apart'. It is also useful to remember that the bubble always moves in the same direction as the *left* thumb.

When the bubble is central, the telescope is turned through 90° to lie over the third foot screw. Again the bubble is centred, but using only this third screw (Fig. 5.8(b)).

The telescope is then turned back through the same quadrant to return it to the first position. It is more than likely that the bubble will have moved slightly off centre and require to be recentred.

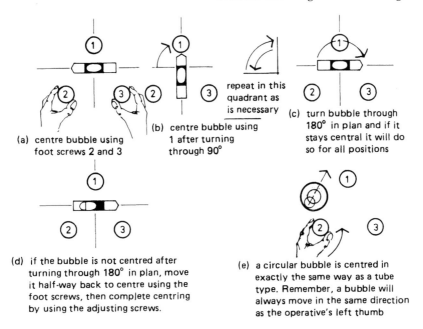

(a) centre bubble using foot screws 2 and 3

(b) centre bubble using 1 after turning through 90°

repeat in this quadrant as is necessary

(c) turn bubble through 180° in plan and if it stays central it will do so for all positions

(d) if the bubble is not centred after turning through 180° in plan, move it half-way back to centre using the foot screws, then complete centring by using the adjusting screws.

(e) a circular bubble is centred in exactly the same way as a tube type. Remember, a bubble will always move in the same direction as the operative's left thumb

Fig. 5.8 Levelling-up

The telescope is moved backwards and forwards through this same quadrant in plan until the bubble stays centred in both positions. Once this has been achieved, the telescope is turned through 180° in plan – i.e. reverse ends (Fig. 5.8(c)). If the bubble stays centred, it will remain so no matter in what direction the telescope is pointed and is said to *traverse*. If the bubble does not traverse (Fig. 5.8(d)), then the *permanent adjustment* of the level is faulty (see Section 5.20). Where the maladjustment is only slight, it is still possible to use the level provided that the bubble is centred before every observation is taken, in much the same way as the tilting level is adjusted immediately before reading the staff.

b) The tilting level The level will have a circular bubble which is centred by a method which depends on its mounting.

   i) *Three foot screws* There is no need to rotate the instrument, and the bubble is centred by appropriate movements of the three screws until it is in the centre of the ring on the bubble glass (Fig. 5.8(e)).
   ii) *Ball-and-socket mounting* Holding the instrument firmly with the left hand, loosen the joint with the right hand and tilt the instrument in various directions until the bubble is centred. Take care when reclamping the mounting to use a firm but steady motion, since a jerky movement will put the bubble off-centre.

c) The automatic level All automatic levels will have a circular bubble, with the lower-accuracy instruments having a ball-and-socket mounting. Higher-accuracy levels will have a three-foot-screw system, and in both cases the centring is as described for the tilting level.

5.5.3 Eliminating parallax

This was mentioned briefly in Chapter 4 and is the same for all telescope instruments. The telescope is racked out of focus and pointed at a light background (e.g. the sky is ideal if it can be found in the horizontal plane). Screw the eyepiece right in clockwise; then, looking through the telescope, slowly screw the eyepiece out again until the reticule pattern is sharp and black.

In order to check, focus the telescope on some distant object and move the eye up and down. There should be no relative movement between the object and the cross-hairs. If there is, adjust until they are 'glued' together. If the eyepiece is fitted with a *diopter scale*, once parallax has been eliminated the setting should be the same each time a particular surveyor uses that instrument. The diopter scale indicates how far the eye-piece has been screwed in or out in relation to the centre of its run and is marked on the outer casing of the eyepiece to match graduations on the instrument itself.

5.5.4 Observation of the staff

Once again the procedure will vary according to the type of level being used.

a) The dumpy level   The telescope is aimed at the staff by sighting over the top of the telescope, which may or may not be fitted with rifle sights for this purpose. Once the staff has been found, the telescope is focussed and turned until the *vertical cross-hair* bisects the middle of the staff. Final adjustments to the aim can be achieved by means of the slow-motion screw (if fitted).

Check that the bubble is still centred – if not, adjust using the appropriate foot screw – then read the graduation on the staff which is cut by the horizontal cross-hair (see Fig. 4.8). Note this reading in the level book, then re-observe the staff as a check.

b) The tilting level   The operation is exactly as described for the dumpy level except that *the bubble is centred by means of the tilting screw immediately before reading the graduation on the staff*. This *must* be done immediately before *every* reading, and in practice this adjustment makes little or no difference to the height of collimation and takes only a second or so to carry out.

c) The automatic level   Again the operation is exactly as previously described – aim, focus, and bisect the staff. However, before each observation, while looking through the telescope, tap gently either the telescope or the tripod leg to see that the prism mechanism is working freely. The horizontal cross-hair should move slightly but stabilise quickly, following which the staff is read and noted in the book.

Whenever a reading is made, the routine is aim, focus, bisect staff, *read, book, and read again (check)*. If this simple procedure is followed, accurate results will surely be obtained.

5.5.5 Moving the level to a new position

There is no need to detach the level from the tripod in order to move it, provided that the instrument is securely fastened to the tripod and they are carried properly.

If the level has foot screws, these should be turned so that they are central in their run. (No adjustment is required in the case of a ball-and-socket joint.) The tripod is gently lifted so that its legs come together, and the whole apparatus is carefully carried in a near vertical position, resting against one shoulder, to the new position. *Never ever* carry a tripod with instrument attached rifle-fashion, because this does the instrument no good whatsoever.

Once the new position has been reached, the level is set up in exactly the same way as it was at the previous position, by carrying out the temporary adjustments as described above.

One final point. Remember that the level is removed from its case and the tripod is divested of its carrying straps at the start of the field operations. These items, and any other loose equipment, should be locked in the car until the end of the job or until they are needed, or, if the site has been reached by other forms of transport, placed carefully at a base point that can be kept under surveillance at all times. The author still remembers chasing a little rascal who was making off with equipment which had been in full view. It took about half a mile before he was finally caught, which proves that one cannot be too careful and also that a certain standard of fitness is sometimes required for surveying purposes.

5.6 Types of level book

Depending on the purpose for which the levelling is being carried out, various systems of taking observations may be employed. No matter what system is selected, the observations are recorded, along with other relevant information, in a ruled *level book*.

Two standard layouts are readily available, namely

a) **height of collimation** and
b) **rise and fall**.

Individual surveyors may prefer one to the other, but the student should be thoroughly conversant with both rulings. The student will see that the height-of-collimation level book is based on the *method (a)* of calculating the R.L. of a point (Section 5.2.1), while the rise-and-fall book uses *method (b)* (Section 5.2.2).

Both types of layout can be seen in Figs. 5.9 and 5.11, and the student will note that six of the columns are the same for both books.

i) Backsight (B.S.) In which the readings are
ii) Intermediate (I.S.) noted as they are observed
iii) Foresight (F.S.) in the field.

iv) Reduced level (R.L.) In which the known starting or finishing R.L. and those R.L.'s calculated from the observations are written.

| v) Distance | In which other information |
| vi) Remarks | relevant to the operations is written. |

The remaining columns highlight the difference between the two rulings.

| vii) Height of collimation (H. of C.) | In which the height of the instrument is written as it is calculated for each different instrument position. |
| viii) Rise (*R*) ix) Fall (*F*) | In which the rises and the falls are written as they are calculated from the observations of the staff. |

5.6.1 Booking readings

When readings are taken, they are entered on successive lines of the book and *all entries referring to one point on the ground must be entered on the same line of the level book*. There is, however, one exception to this rule and this is at a change point when using the height-of-collimation layout. The reason for this is to avoid confusion when the height of instrument changes. It is far better to write down the foresight and backsight readings on successive lines and to write down the same R.L. for each. This can be seen in Fig. 5.11, where the new height of collimation relates to the backsight which was read to it, while the foresight relates directly to the old height of collimation from which the reduced level of the point was calculated.

The student should purchase both types of book in which to practise exercises and for use in practical work.

5.6.2 Carrying forward level readings

Where observations take up several pages of a level book, it is necessary for the *last reading on any page to be a foresight*, so that the results can be checked before proceeding to the next page where the *first entry must be a backsight*.

If the last entry of the page does not happen to be a foresight but is an *intermediate sight*, it is nevertheless entered in the *foresight column*. This same reading is then entered in the *backsight column of the next page* as the first entry on that page. All other information such as R.L., distance, and remarks will be the same for the 'backsight' entry as for the 'foresight' entry on the preceding page.

5.6.3 Level-book checks

No matter which type of level book is used, an *arithmetical check* should be applied either at the end of the operation or at the end of each page when entries are carried forward over several pages.

The checks are as follows.

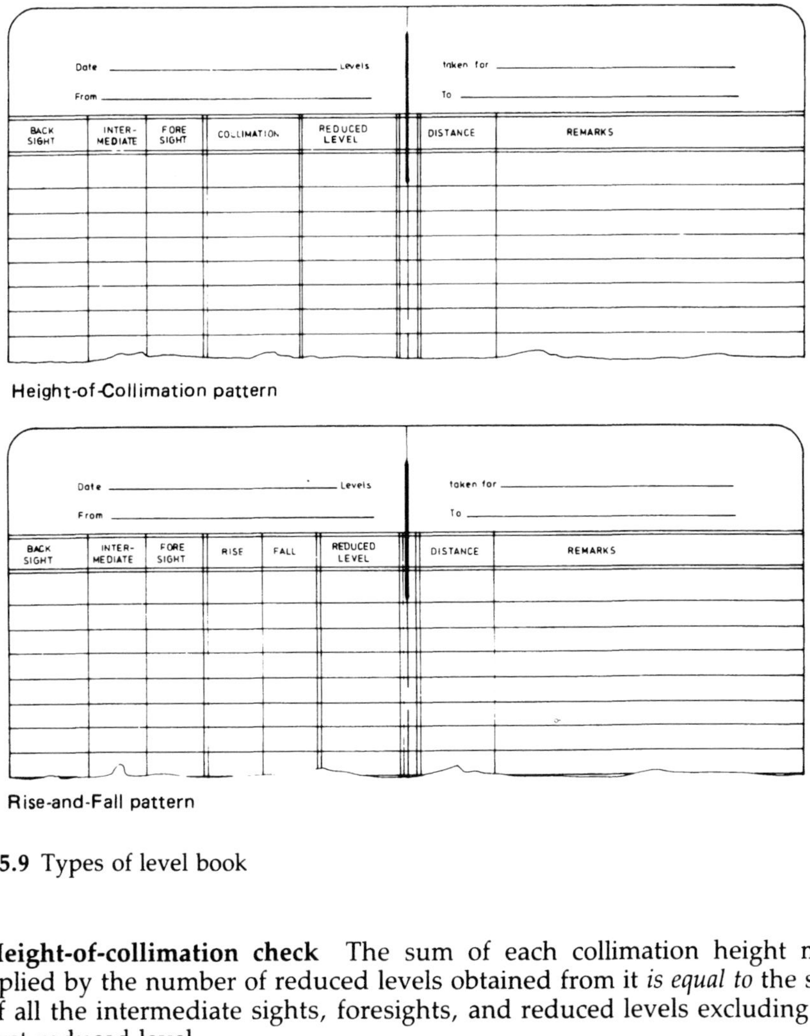

Height-of-Collimation pattern

Rise-and-Fall pattern

Fig. 5.9 Types of level book

a) **Height-of-collimation check** The sum of each collimation height multiplied by the number of reduced levels obtained from it *is equal to* the sum of all the intermediate sights, foresights, and reduced levels excluding the first reduced level.

b) **Rise-and-fall check** The sum of the backsights minus the sum of the foresights *is equal to* the sum of the rises minus the sum of the falls, and *is also equal to* the first reduced level minus the last reduced level.

This can be written mathematically as

$$\Sigma \text{B.S.} - \Sigma \text{F.S.} = \Sigma R - \Sigma F = \text{first R.L.} - \text{last R.L.}$$

Before carrying out the arithmetical check, each page should be examined to make sure that there are the same number of backsight entries as there are foresight entries. It is easy to forget to put the last entry on the page in the foresight column even though it may be an intermediate, or to put a genuine foresight in an intermediate column. When mistakes are found, *never* rub them out – simply put a pencil stroke through the incorrect entry and write the correct figures neatly over the top or in the correct column as the case may be. Always use a pencil with a sharp point when making up a level book. Ballpoint pens, fountain pens, and indelible pencils are not to be recommended if a neat, tidy, and legible book is to be kept.

The check for the rise-and-fall method is somewhat simpler and less tedious than the check for the height-of-collimation method and it is doubtful if the H. of C. check would ever be used in lower-order work. The student must appreciate, however, that the checks *only prove the arithmetic* and that the only way to check the observations is to finish the operations by levelling back on to the starting point, or on to another point of known height.

It is advisable to leave a few lines blank at the bottom of each page, so that a line may be drawn across under which the check figures may be written. Each page will then be taken to a summary on the last page of the set.

5.6.4 Merits of the two types of level book

Comparing the two types of level book, it can be said that the height-of-collimation type is more simple because it uses less arithmetic. On the other hand, the rise-and-fall method can be more completely checked. Both types should be thoroughly understood, and it is the individual preference of the surveyor which will determine which type of level book is used for the task in hand.

All important work is normally carried out by the rise-and-fall method, the height-of-collimation method being adopted only for low-accuracy ordinary building-site work. The beginner is therefore strongly advised to start by using the rise-and-fall method and, if necessary, checking the entire survey by the H. of C. method until such time as he can operate both methods with ease.

5.7 'One-set-up' levelling

This system gets its name from the fact that the level is *set up in only one position*, from which all the required observations of the staff can be made. Figure 5.6 shows the typical case. The level is set up at station I_A, and the staff is observed on successive points A_1 to A_5 *inclusive*. Point A_1 is an O.S.B.M. of known R.L. 250.000 m above O.D. and the R.L.'s at the other points require to be found.

By referring back to Section 5.2, example 2, the student should have no difficulty in relating the process of making up the two types of level book illustrated in Fig. 5.9. Remember, however, that if it is required to check the actual readings, the levelling would have to be taken back to A_1, unless A_5 was of known height.

The 'one-set-up' system is an ideal field use for the dumpy level, which is possibly faster in use than the tilting level, although there is no doubt that the automatic level would be the fastest of all.

5.8 Series levelling

When it is necessary to set up the level in several positions and so work in stages by means of change points, the system employed is called either *series levelling* or *continuous levelling*.

A typical example is shown in Fig. 5.10, where, because of obstacles and steepness of slope, *four* instrument positions – I_A, I_B, I_C, and I_D – are necessary. The ground points where the staff is held and observed are noted as A_1 to A_4 (from station I_A), B_1 to B_4 (from I_B), C_1 to C_3 (from I_C), and D_1 to D_3 (from I_D). The student should note which are backsights, intersights, and foresights.

Points A_1 to A_4 are entered into the level book (see Fig. 5.11) *in exactly the same way as they were in 'one-set-up' levelling*. Point A_4, however, is now a change point, because this is the last point upon which an observation will be made from I_A before the instrument is moved to its new position I_B. *The first reading, B_1, will therefore be on the same point of ground as A_4 but read to a different height of collimation.* Although the two readings will have different values, because they are the same point of ground *the R.L. will remain the same* and the student should note how the readings are entered into the level book:

a) in the rise-and-fall book, they are entered *on the same line*, with A_4 written in the *foresight* column and B_1 written in the *backsight* column;
b) in the height-of-collimation book, they are written on *successive lines*, A_4 still in the *foresight* column and B_1 in the *backsight* column, with the R.L. written *on both lines*. Note, however, that the new height of collimation is written on the same line as B_1 (backsight).

All the readings taken from I_B will then be entered into the level book as though they were also for a 'one-set-up' situation. This will also apply to the readings from station I_C and I_D, with all change points noted in the same manner as A_4/B_1.

In effect, what we have is *a series of 'one-set-up' situations* connected by the respective change points at A_4/B_1, B_4/C_1, and C_3/D_1. The reading to A_1 is the first in the series, and the reading to D_3 is the last.

The student should have no difficulty in understanding the pages of the level book as illustrated in Fig. 5.11 and should note how the checks have been carried out. Again, if the observations are to be checked, it is necessary to level back to A_1 or to some other point of known height.

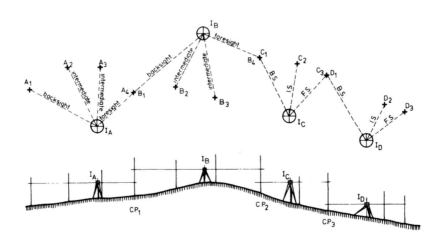

Fig. 5.10 Series levelling

Level book — Collimation (height of instrument) method

Date 28 May 1981 From A₁ (Bench Mark) To D₃ (Manhole)   Riverside Farm   Levels

BACK SIGHT	INTER-MEDIATE	FORE SIGHT	COLLIMATION	REDUCED LEVEL	DISTANCE	REMARKS
2.400			152.400	150.000		A₁ (B.M. Start)
	2.000		"	150.400		A₂
	1.900		"	150.500		A₃
2.800		1.400	153.800	151.000		A₄/B₁ (C.P.1)
	2.000		"	151.800		B₂
	1.400		"	152.400		B₃
1.300		2.600	152.500	151.200		B₄/C₁ (C.P.2)
	2.500		"	150.000		C₂
0.600		3.000	150.100	149.500		C₃/D₁ (C.P.3)
	1.700		"	148.400		D₂
		2.100		148.000		D₃ (END)
7.100	11.500	9.100		150.000		
		7.600		2.000✓		
		2.000✓				CHECKS

Sum of all intermediate sights = 11.500
"   "   foresights = 9.100
"   "   R.L.s (except first) = 1503.200
= 1523.800

Sum of each collimation height × number of reduced levels from it = 1523.800 ✓

Level book — Rise and fall method

Date 28 May 1981 From A₁ (Bench Mark) To D₃ (Manhole)   Riverside Farm   Levels

BACK SIGHT	INTER-MEDIATE	FORE SIGHT	RISE	FALL	REDUCED LEVEL	DISTANCE	REMARKS
2.400					150.000		A₁ (B.M. Start)
	2.000		0.400		150.400		A₂
	1.900		0.100		150.500		A₃
2.800		1.400	0.500		151.000		A₄/B₁ (C.P.1)
	2.000		0.800		151.800		B₂
	1.400		0.600		152.400		B₃
1.300		2.600		1.200	151.200		B₄/C₁ (C.P.2)
	2.500			1.200	150.000		C₂
0.600		3.000		0.500	149.500		C₃/D₁ (C.P.3)
	1.700			1.100	148.400		D₂
		2.100		0.400	148.000		D₃ (END)
7.100		9.100	2.400	4.400	150.000		
		7.600		2.400	148.000		CHECKS
		2.000✓		2.000✓	2.000✓		

Fig. 5.11 Pages of a level book relating to Fig. 5.10

5.9 Reciprocal levelling

This is a system of levelling whereby the difference of height between two points at a considerable distance apart can be accurately determined. All errors due to the curvature of the earth, refraction by the atmosphere, and faulty adjustment of the level are in fact, for all practical purposes, eliminated.

Figure 5.12 shows a wide river with two points A and B whose difference in height needs to be determined. The level is set up at I_1 and, with the staff on A and B, readings are taken as a_1 and b_1. The level is then set up at I_2 and, with the staff on B and A, the readings are taken as b_2 and a_2. (Note that the distances I_1A and I_2B should be equal.)

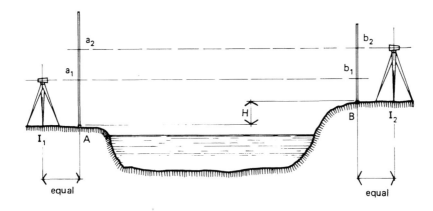

Fig. 5.12 Reciprocal levelling

Let H be the difference in height between A and B, and $\pm x$ be the error in the distance AB due to curvature, refraction, and incorrect adjustment; then

$$H = a_1 - (b_1 \pm x) \tag{i}$$

also $\quad H = (a_2 \pm x) - b_2 \tag{ii}$

hence $\quad 2H = (a_1 - b_1) + (a_2 - b_2)$

$\therefore \qquad H = \dfrac{(a_1 - b_1) + (a_2 - b_2)}{2}$

This proves that the true difference of height is the *mean* of the two determinations. (The student should note that when adding the equations (i) and (ii) together to arrive at the value of $2H$, the error factors of $\pm x$ cancel each other out.)

Example In levelling across a river with the level in the first position (I_1) the staff readings were 1.200 m at A and 0.600 m at B. In the second position, the readings were 2.000 m at A and 1.450 m at B. What is the true difference of height between A and B?

Let H represent the required difference in height,

$$\therefore H = \frac{(a_1 - b_1) + (a_2 - b_2)}{2}$$

$$= \frac{(1.200 \text{ m} - 0.600 \text{ m}) + (2.000 \text{ m} - 1.450 \text{ m})}{2}$$

$$= \frac{0.600 \text{ m} + 0.550 \text{ m}}{2} = \frac{1.150 \text{ m}}{2}$$

$$= 0.575 \text{ m}$$

It should be noted that in the above method the refraction is presumed to be constant while the readings are being taken. By using two levels and two staves the readings may be taken simultaneously, in which case the error due to refraction will be eliminated completely.

One of the difficulties of using two ordinary levels is to get them adjusted so as to be in perfect agreement for collimation, otherwise an error will be introduced. This can be overcome by interchanging the levels and again finding the difference in height. Additionally, equipment such as 'valley-crossing equipment' has been developed to deal with this type of problem and incorporates two similar instruments in correct adjustment.

Where it is impracticable to carry out reciprocal levelling or to equalise sight lengths, high-accuracy operations would require corrections for curvature and refraction to be applied to the staff readings. The curvature of the earth causes staff readings to be too large, while the effect of refraction is to curve the sight line downwards from the horizontal. For average conditions the correction to a staff reading to eliminate both types of error may be taken as $-(K^2/15)$ m, where K is the sight length in kilometres. It is highly unlikely that the student will use corrections of this order in straightforward ordinary levelling, but it is as well to be aware that they exist as a preparation to later more advanced studies.

5.10 Flying levels

These are readings taken for the purpose of checking a series of levels already taken in greater detail. Taking *flying levels* is, in effect, a system of levelling without intersights, where every point of ground levelled is treated as a change point. The system is used when levelling between two points such as two B.M.'s or from B.M. to T.B.M. which are well apart from each other in terms of either distance or height.

It has been shown that in ordinary levelling there is no field check on the intersights, and any closing error will be created by B.S. and F.S. errors at the change points. It follows, therefore, that the closing error reflects the exact value of the fieldwork, and that is the reason for levelling back either through the change points already used or via new ones, by means of flying levels, to finish the levelling on the starting point or on some other point of known height. In order to reduce errors to a minimum, the backsight and foresight distances should be kept equal.

5.11 Trial levels

These are simply a short series of levels taken merely to determine the difference in height between two points, possibly as a reconnaissance before a more detailed survey is carried out.

5.12 Principles of area levelling and contouring

Area levelling can range between taking a few *spot heights* or *spot levels* over a site, for general use, to taking a dense network of levels for the purpose of determining *contours*.

5.12.1 Definitions

Some definitions are necessary as an introduction to the subject.

a) 'Spot heights' or *'spot levels'* is the term used for levels on a site which are taken in a random manner on various 'spots' or points of ground. Small building-site surveys can provide sufficient height information from a few levels scattered over the site to enable minor drain layouts etc. to be worked out without the need for contours.

Spot levels are usually observed from a one-set-up situation and, such is the low accuracy required for simple building operations, readings need be taken only to the nearest 5 mm or even 10 mm.

Where levels have some importance, e.g. in setting up T.B.M.'s or structural height lines, even though they may be spot levels, the levelling should be carried out carefully and be read to 1 mm.

In the author's opinion, it is easier to get into the habit of reading to 1 mm no matter what the task and to carry out any adjustments in the office *after the R.L.'s have been calculated*.

b) A contour line or *contour* is an imaginary line on the surface of the earth, every point on which is at the same height or altitude. The best example of such a line is the water's edge of a still lake.

When drawn on a map or plan, a series of such lines, of differing heights above the datum, will provide useful information for purposes of designing layouts for roads, railways, drainage schemes, and other engineering and building operations.

Figure 5.13 shows a contour plan and a typical section. The student should note how the actual height of each contour is written within its length, the convention being that the tops of the figures are towards the higher slopes or the summit. The actual peak or summit of a hill is usually marked by a small dot with its spot height written alongside.

c) The vertical interval (V.I.) is the term used to denote the difference in height between successive contour lines and it is usually constant on any one drawing. The V.I. used on a map or plan will depend on

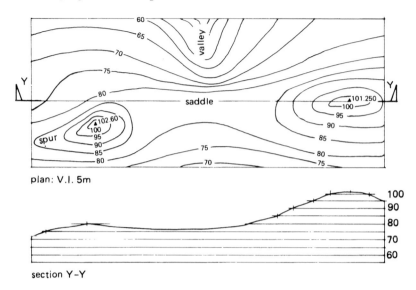

plan: V.I. 5m

section Y–Y

Fig. 5.13 Contours

    i) the scale of the drawing,
    ii) the purpose for which the drawing has been prepared,
    iii) the nature of the ground.

Large-scale drawings such as are used for engineering and building operations may use V.I.'s from 0.5 m to 3.0 m. Maps for topographical purposes may have a V.I. from 3.0 m to 15.0 m, while mountainous regions such as the Swiss Alps often use a V.I. as large as 30 m. (Note: The old 'one inch' O.S. maps of Great Britain showed contours at 50 ft intervals; the contours on the new metric editions are the same but are noted to the nearest metre.)

d) The horizontal equivalent (H.E.) is the term used to denote the *shortest* horizontal distance between successive contour lines. The H.E. will vary with the slope of the ground (see Section 5.12.2).

e) The gradient is the slope of the ground as determined by combining (c) and (d),

$$\text{i.e. gradient} = \frac{\text{vertical equivalent}}{\text{horizontal equivalent}} = \frac{\text{V.I.}}{\text{H.E.}}$$

The gradient is usually expressed as a ratio of the height risen in a horizontal distance travelled, e.g. to travel a horizontal distance of 10 units in order to rise 1 unit would be written as 1 in 10 or expressed as a percentage, i.e. 10%.

5.12.2 Reading a contour map or plan

The important features of an area of land can readily be seen from the contours as follows.

a) The direction of the steepest slope is always at right angles to the contour line. A proposed route may therefore have to cut the run of the contours on a diagonal line in order to maintain a reasonable gradient.
b) Contour lines which are close together indicate a steep slope, i.e. a small H.E. indicates a large gradient.
c) Contour lines which are widely spaced indicate a gentle slope.
d) A contour 'island' indicates either a *hill* or a *depression*, according to how the levels are changing.
e) Two 'islands' close together will indicate either

   i) two hills with a 'pass' between them, or
   ii) two depressions with a 'ridge' between.

f) Contour lines *can never cross*.
g) Contour lines must be continuous. A single line cannot split into two lines each having the same value.
h) Only in the case of a vertical cliff can contour lines join.
i) A contour line cannot simply end – it must close back on itself, though not necessarily on any one map.
j) A peak or summit will be indicated by a small dot with the relevant spot height alongside. If it is also a *triangulation point*, the dot will be enclosed in a small triangle. Triangulation points are used by the Ordnance Survey as control points for the mapping of the UK and are indicated as described on O.S. maps and plans.

5.12.3 Use of a contour map or plan

A contour map or plan has many uses. It may be used in reconnaissance work in order to plan a preliminary route; to draw to scale approximate ground profiles or sections to check intervisibility along any line on the plan; to calculate approximate volumes of earthworks; to calculate the approximate capacity of reservoirs; and so on – all of which can form a basis upon which discussions can take place in order to arrive at decisions which will affect later detailed work.

5.13 Direct levelling for contours

Direct contouring is a method whereby the contour is found by means of ordinary levelling technique, each point on the contour line being pegged out in the field. This is sometimes termed *pegging-in contours*, and the resulting contours are *surveyed contours* or *instrument contours*. It is a twofold operation, used for locating contours exactly in the field, and results in a minimum of office work in terms of plotting the contours on the survey drawing.

5.13.1 Surveying the pegged line

There are two ways by which the pegs can be surveyed once the contours have been located on the ground:

i) by means of standard horizontal-measuring equipment, e.g. chain and tape. It is perhaps better to locate the contours and place the pegs *before* the chain survey is carried out. This will obviously save time because the pegs can be picked up as part of the normal survey detail, e.g. using offsets, ties, etc. Where no separate chain survey is in prospect, the measurement of the pegged lines would form its own survey.

ii) by means of *tacheometry* (see Section 5.3), which is faster than (i) but requires a tilting level with a horizontal circle of degrees to be used for locating the contours.

(Note: A traverse may be employed as a third method of surveying the pegged line. This would be more appropriate in difficult terrain, but would be the subject of more advanced studies.)

5.13.2 Field routine

i) Levelling is begun from any convenient B.M. and is continued up or down hill until a point roughly on the contour has been found.

ii) The height-of-collimation method of booking must be used, and the height of collimation of each set-up of the level has to be determined before moving the instrument to a new station.

iii) When the H. of C. is between 0.5 m and 3.0 m above the required contour, levelling is halted.

Example It is required to peg out the 100 m contour, and the levelling has reached the stage where the H. of C. is 101.525 m. It is obvious from the system of calculating R.L.'s by the height-of-collimation method that, when the staff reading is 1.525 m at the centre cross-hair of the level, the foot of the staff will be on the required contour line,

i.e.   H. of C. − staff reading = R.L.

∴   101.525 m − staff reading = 100.000 m

∴   required staff reading = 101.525 m − 100.000 m

$$= 1.525 \text{ m}$$

It follows then that, while the instrument remains in position, the foot of the staff will be on the required contour line each time the staff is observed to read 1.525 m.

iv) A peg is inserted at the point and is marked with the contour value.

v) To simplify the operation, the staff may have the required reading marked with a strip of tape, or a target may be used as a substitute. The target is held vertically on the contour and is adjusted to the true horizontal line of sight. In both cases, the observer will have the easier task of looking for a simple mark as opposed to one graduation among many on the surface of a normal staff.

vi) The staffman now moves about twenty paces along what he estimates to be the contour and holds the target vertically on the ground. The observer waves him up and down the slope until the target is intersected by the line of collimation at the required mark. Again a peg is inserted.

This operation continues until all points in the required contour line have been pegged in.

vii) When the target is approximately 30 m beyond the level, a change point is made – but not in quite the same manner as in ordinary levelling. The target remains where it is on the contour, while the level is set up at its new position about 30 m beyond the target. Instead of reading a B.S., the target is adjusted to the new height of collimation and the pegging is resumed up to and beyond the instrument. This is repeated until the required contour is complete.

viii) The operation should always be continued to abreast of a B.M., and the difference in height between the B.M. and the last peg should be found by ordinary levelling as a check.

ix) The line of pegs will then be surveyed.

Where the surveying is done by means of *tacheometry* using a tilting level (see Section 5.3), the staff marked with a strip of tape is used in preference to the target, and the following changes to the routine occur.

v') The staff will have the required reading marked with a strip of tape. A reference direction for bearing is selected and the horizontal circle of degrees is orientated accordingly. This direction should be used for all instrument positions.

The telescope is aimed at the staff, held vertically on the contour line.

Using the tilting screw, the *bottom stadia line* is brought to an exact 0.1 m (decimetre) graduation. The value of this is booked down. The value of the *upper stadia line* is read and booked. The difference between the two is then multiplied by the constant of the instrument (usually 100) and this is also booked down as the distance from the instrument to the staff.

Example When the centre cross-hair was reading 1.525, the lower stadia line read 1.375. This was moved to 1.400, and the upper stadia line then showed 1.650. The difference is $1.650 - 1.400 = 0.250$, which when multiplied by $100 = 25$ m (distance from instrument to staff).

Finally, the horizontal bearing on the circle of degrees is noted and booked.

vi') The staffman moves about twenty paces along what he estimates to be the contour and holds the staff vertically on the ground. The observer waves him up and down the slope until the staff is intersected by the centre cross-hair at the required mark. A peg is inserted in the normal way.

The operations in (v') and (vi') continue until all the points in the contour line have been pegged in.

vii') When the staff is about 30 m beyond the level, a change point is made as follows. The staff is held on the peg while the level is moved to, and set up at, its new position. A backsight reading is taken and a new height of collimation is calculated, from which the new required staff reading is determined. The staff is then 'taped' at this new required reading, following which the operations are continued in the manner described.

5.13.3 Summary

To reduce the likelihood of error, it is simpler and easier to peg all the points in any one contour, where several contours can be located from one instrument position, locating each contour in turn. In the example shown, with the height of collimation at 101.525 m and using a 4 m staff, contours at 98 m, 99 m, 100 m, and 101 m could be located. Each would be done separately from the one staff position, as far as the range of the instrument would allow, and, when it became necessary to move the instrument to a new position, this position would be selected so that the same range of contours could be located from the new line of collimation.

When peg positions are to be surveyed tacheometrically, the site positions of the instrument must be fixed on plan to enable the plotting process to be carried out in the office. Once the instrument stations and the reference direction have been drawn, the directions from each station to the various staff positions can be set out by protractor and the calculated stadia distance can be scaled off along these lines to give the points on the contour line. These points are then joined together by a smooth freehand curve to represent the contour line.

The student may recall from Chapter 1 that plotting by angle direction and distance is termed 'polar co-ordinates' or 'radiation'.

5.14 Indirect levelling for contours

Indirect contouring is a method whereby lines are ranged out on the ground and the R.L.'s of points along the lines, usually at regular intervals, are found by ordinary levelling techniques. The positions where the contours cut the lines are calculated from the observed R.L.'s by means of *interpolation* (see Section 5.14.5) which is done after the plotting of lines on the survey drawing. Once the contour positions have been determined, they are joined together by smooth curved lines to complete the drawing. Although the spot levels are put on the drawing initially, once interpolation is complete they are rubbed out or left on, to choice.

There are several ways by which height information may be obtained in the field, and the principal methods are described below.

5.14.1 Grid levelling

The most systematic, and probably the most commonly used, method of obtaining the information for indirect contouring is *grid levelling*. Several operations are involved:

   i) the area must first be surveyed;
  ii) the area is then marked out with a grid;

iii) spot levels are taken at the points of intersection of the grid lines;

iv) the grid is plotted on the survey drawing and the contours are found by interpolation.

When undertaking a full survey of, say, an open field, the grid is set out from the longest chain line in such a way that it can be plotted accurately on the survey drawing or an overlay. The size of the grid will depend on the slope of the ground. For gentle slopes, the squares of the grid can be 20 m, but 5 m or 10 m may well be used for steeper slopes. A combination of both is not uncommon where the overall site may be reasonably gentle in its gradient but have areas of local steepness which require a denser network of spot levels to be taken. It may also be necessary to take occasional spot levels inside a square or off the grid, in which case the points must again be surveyed so that their positions can be plotted accurately.

The student must remember from Chapter 2 what was said about *horizontal measurement*. No matter what type of grid is laid out on the ground, it will be plotted to scale on a plan, and therefore the grid needs to be set out by means of *horizontal* measurement and *not on the slope*.

The grid itself may be set out in a variety of ways, but the most sensible method is to place a double line of ranging rods along each of two adjacent sides of the grid (see Fig. 5.14(a)). It is then a simple matter for the staffman to line himself in by means of pairs of rods in order to place the staff at the intersections of the grid lines where levels are required. On a small site, it is not uncommon for the whole of the grid to be marked out. This should be done by means of alternate ranging poles and arrows in both directions, which will result in diagonal lines of poles and diagonal lines of arrows (see Fig. 5.14(b)).

Lines in one direction are given letters – A, B, C, D, etc. – while lines in the other direction are given numbers – 1, 2, 3, 4, etc. This will ensure that every point of intersection can be specified by its co-ordinates, e.g. B6 is the intersection of lines B and 6.

The levelling is carried out by the ordinary levelling process, working from the nearest B.M. when absolute heights are required. If relative heights will suffice, an arbitrary point may be selected as datum. The levelling should be systematic – that is to say, up one line and down the next in a continuous pattern (see Fig. 5.14) – and, if the site is small, about 100 m square maximum, and reasonably flat, one set-up of the instrument should be sufficient. If change points are necessary, the normal procedure for these should be followed.

The booking, by either the height-of-collimation or the rise-and-fall method, must be done with care so that each reading relates to a co-ordinate. When the last point of the grid has been reached, a return to the starting point must be made by means of *flying levels*. If only one set-up has been necessary, the check is to reobserve the starting peg to ensure that, in the case of the dumpy level, the line of collimation has not altered.

5.14.2 Contouring by sections

Following the method of section levelling, a long line may be ranged out through the area and sections of levels be taken *left and right* of the longitudinal line (see Fig. 5.14). The spot levels may be taken either at uniform

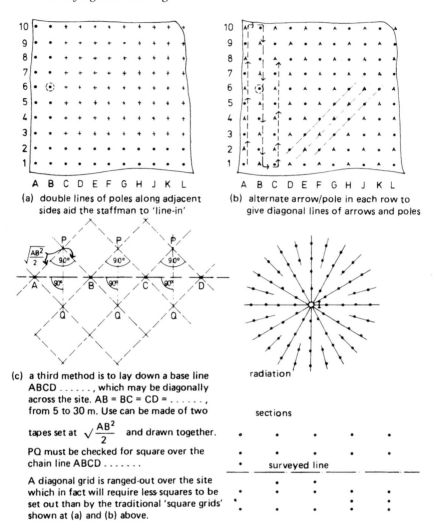

(a) double lines of poles along adjacent sides aid the staffman to 'line-in'

(b) alternate arrow/pole in each row to give diagonal lines of arrows and poles

(c) a third method is to lay down a base line ABCD ......, which may be diagonally across the site. AB = BC = CD = ..., from 5 to 30 m. Use can be made of two tapes set at $\sqrt{\dfrac{AB^2}{2}}$ and drawn together. PQ must be checked for square over the chain line ABCD ....

A diagonal grid is ranged-out over the site which in fact will require less squares to be set out than by the traditional 'square grids' shown at (a) and (b) above.

radiation

sections

surveyed line

Fig. 5.14 Grid-levelling patterns

distances from the centre line or only at points of change of ground slope. The points taken are plotted on the plan and the contours are interpolated between them in the normal way.

5.14.3 Contouring by radiating lines

The level, with a horizontal circle of degrees, is set up over a fixed known spot in the centre of the area under consideration. A reference direction is chosen, and levels and distances on several lines radiating from the instrument position are observed. As the radiating lines open out, the gaps are filled by additional lines in order to increase the density of levels taken, the result being a cobweb pattern as can be seen from Fig 5.14. It is a useful method on a small hilltop or knoll to enable the slopes of the hillside to be contoured, and the author has adopted the method in lowland situations to good effect.

5.14.4 Contouring by tacheometry

This is the fastest of all the field methods, providing a tilting level fitted with a horizontal circle of degrees is used. The levels can be taken where they will best reflect the nature of the ground, rather than in some pre-determined pattern of, say, a grid. Observations to non-regular points can considerably increase the fieldwork of other methods – notably grid levelling. Tacheometric levelling therefore avoids this extra work.

Once again the level is set up over an accurately located point of ground, which may be central on a small site, and a reference direction is selected. The levelling is started from a B.M. in the normal way, and the observer directs the staffman to hold the staff at the most suitable ground points.

The routine is the same as that used for direct levelling when tacheometry is used:

i) Aim on the staff, focus, and bisect by means of the vertical hair.
ii) Centre the bubble and read at the centre cross-hair. Book and read again as a check.
iii) Using the tilting screw, bring the lower stadia hair to an exact decimetre graduation on the staff, and book.
iv) Read the *upper* cross-hair, book, and reread (a check). Carry out the arithmetic and book the calculated distance to the staff.
v) Read and book the bearing on the horizontal circle of degrees.
vi) Direct the staffman to new position and repeat the routine starting from stage (i).

Again the points are plotted by means of polar co-ordinates using a protractor and a scale rule, with interpolation being used to determine the contours.

Note: Where tacheometric principles are to be employed, the dumpy level and the automatic level are not suitable, since an exact decimetre cannot be obtained at the lower stadia line. A greater error is induced when this cannot be done, because two readings are estimated and not just the one at the upper stadia line, as is the case when the tilting level is used.

5.14.5 Interpolating contours

Although time-consuming in the office, *interpolation* is a relatively simple mathematical process. No matter which method is used in the field, the ground points are plotted on the plan in their correct positions and the spot level is written alongside each point. Since the contours will be sketched in freehand, the spot heights may be written on the drawing to 0.1 m or 0.01 m, even though readings are taken to three figures in the field.

Figure 5.15 shows a portion of a grid of levels. The method is to plot the points of intersection of each contour with the grid lines, then draw in the contours as smooth curves.

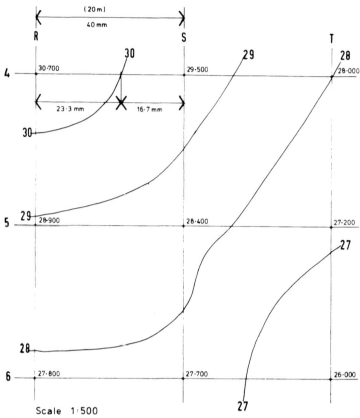

Fig. 5.15 Interpolation of contours

a) The grid interval on site is 20 m and, when drawn to a scale of 1:500, results in a *drawn grid interval of 40 mm*.

b) Along line 4, between R4 and S4, the ground falls from 30.700 m to 29.500 m; therefore, along this section 40 mm represents a fall of 1.200 m.

c) From spot level 30.700 m (R4) to the point where the 30 m contour crosses grid line 4, the fall is 30.700 m − 30.000 m = 0.700 m.

d) A fall of 1 m is represented on the drawing by (40/1.2) mm (from (b)); therefore the distance from R4 to the 30 m contour will measure on the drawing 0.7 × (40/1.2) mm, i.e. 23.3 mm.

   This can be checked by calculating the corresponding distance from S4 to the 30 m contour line. The ground rises 0.5 m, therefore the distance will measure on the drawing 0.5 × (40/1.2) mm, i.e. 16.7 mm. When the two distances are added together (23.3 mm + 16.7 mm) the result is 40 mm, which is the actual distance between R4 and S4 as measured on the drawing.

e) The same procedure is followed in order to plot all other contour/grid intersection points by considering spot heights.

f) Once all the points have been found, *all corresponding points are joined by a smooth freehand curve as shown*.

There is a further method of interpolation by means of an *interpolation graph*, but its use is rather tedious, since the operative has to work underneath a sheet of tracing paper on which the grid pattern has been drawn. The process also necessitates drawing a graph for each job. In the author's opinion, the method of interpolation as described is quite adequate and yields quite satisfactory results once the operative has got used to it.

5.14.6 Summary

The advantage of using a grid is that the average level of the site can be determined quite readily, which is useful when attempting to work out cut and fill relative to earthworks, or to select a possible floor level for a new building.

   Against this can be set the disadvantage of the time taken to set out the grid pattern on site, which can be lengthy, although on small sites this is rather minor. The main disadvantage is possibly the fact that the grid method assumes a uniform slope between any adjoining pair of grid intersections, ignoring any changes in level between the squares. Where major changes to the 'normal' slope occur, this can be overcome by taking additional levels within the squares, although in practice small changes between intersection points have little effect on the accuracy of the final result. (In built-up areas or in woodland, it may not be possible to set up a grid, therefore other ways of carrying out the survey work would be devised and used.)

   A large number of levels are also observed simply to complete the grid, and some of these may not be required for the contouring. This is especially so in areas of gentle slope or smooth relief, although, in practice, it is surprising how often so-called superfluous information can come to the rescue when a problem arises.

   In general terms, contouring can be interesting and relatively simple to carry out and, provided that the usual precautions against error are taken, the results will be satisfactory. By now, the student should not need reminding that others will base their work upon the results of the survey – so it *must* be accurate or their work will be abortive.

5.14.7 Comparison of direct and indirect contouring

a) **Direct contouring**   This is the most accurate method, although a great deal of fieldwork will be required after the levels have been taken – unless tacheometry is used.
b) **Indirect contouring**   The levelling of spot heights by tacheometry is the most convenient method (and is fast). Grid levelling is the most useful, especially when all points of intersection are pegged, although setting out the grid can be laborious.

5.15 Section levelling

In both building and engineering work, it is often necessary to prepare profiles of the ground along a specific line. Termed a *section of levels*, the profile is obtained by taking a number of levels along the line and plotting the heights and distances to appropriate scales.

Unlike pure 'building construction' sections, engineering sections for roads, sewers, etc. are often drawn with the scale for the horizontal distance different from that used for the vertical height. The reason for this is that, for a *longitudinal* section, the horizontal distance can be quite long and, in order to reduce the length of the drawing sheet, the scale for the horizontal distance would be smaller than for the vertical heights; e.g. horizontal 1:500, vertical 1:100, which would also emphasise the height variations. For cross-sections the scale may well be the same.

Figure 5.16 shows typical longitudinal sections and cross-sections.

5.15.1 Longitudinal sections

Levels are observed along the required line by means of series levelling, which has already been described. Observations are made with the staff at regular intervals of horizontal distance (e.g. 20, 25, 50, 100 m, etc.) together with the levels of any points where the profile is disturbed, such as at a change of ground slope, obstructions, etc. Work should be started and finished on a B.M. or T.B.M. or be closed by means of flying levels in the normal way, with the length of the backsights and foresights kept equal.

Note in the figure how the ground line is drawn by means of straight lines between the ground points. There is no point in using smooth curves where one of the scales is larger than the other. All the relevant information is shown in boxes drawn across the bottom of the section, and additional empty boxes may be provided to receive further information at a later date.

5.15.2 Cross-sections

If the proposed construction work is of some width, the longitudinal sections will need to be supplemented by *cross-sections*. Such sections are *at right angles* to the longitudinal profile line and are used primarily for the calculation of the volume of earthworks.

The cross-sections are taken at regular intervals from 20 m on broken ground to perhaps 100 m where the slope is gentle. But again, any major changes in profile will need to be picked up, by either additional levels or extra sections. Normal series levelling as discussed in Section 5.8 is used for this work, and it may be done in concert with the longitudinal section levelling or completely separate from it. If the longitudinal line is pegged at the points along its length where the cross-sections are to be taken, then the cross-section levels can prove the accuracy of the longitudinal work.

Although one cross-section is normally completed at a time, then on to the next, in difficult terrain it may be quicker to complete the downhill levels for two or more cross-sections from one set-up and the uphill levels from another. Booking, in this case, needs to be done with extreme care, so that the levels of the different sections are not mixed up.

Finally, when plotting cross-sections, it is usual to use the same scale for both horizontal distance and vertical height. Where sections are done in conjunction with a longitudinal section, the scale chosen would be that used for the vertical height on the longitudinal section.

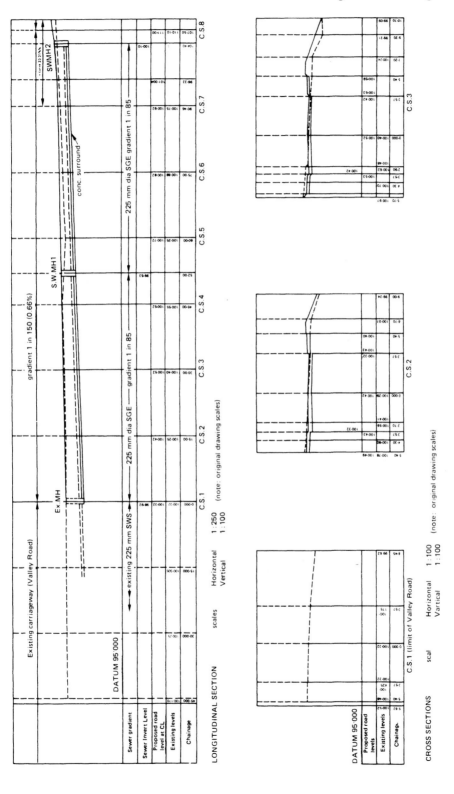

Fig. 5.16 Typical sections

5.16 Inverse levelling

Normally the levels that interest the surveyor are below the line of collimation, but this is not always the case. For example, when levelling through a tunnel, or under a bridge, it may be necessary to check the level of the underside of the roof. This would apply equally to the underside of beams, ceilings, canopies, soffits of arches, etc., in order to check the setting-out, or to continue levelling operations over an obstruction.

5.16.1 Inverted staff position

To check the distance from, say, the underside of a tunnel roof to the line of collimation (see Fig. 5.17), the staff is turned upside-down and the foot is placed up to the roof. The staff is then read and the distance to the line of collimation is booked in the normal way but *prefixed by a negative sign*.

B.S.	I.S.	F.S.	H. of C.	R.L.	Dist.	Remarks
1·275			24·793	23·518		Point A
	−2·447		"	27·240		B (underside Tunnel roof)
		2·853	"	21·940		point C (c.l. road)

B.S.	I.S.	F.S.	R	F	R.L.	Dist.	Remarks
1·275					23·518		Point A
	−2·447		3·722		27·240		B (underside Tunnel roof)
		2·853		5·300	21·940		point C (C.L. road)

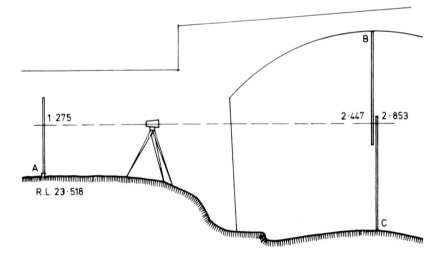

Fig. 5.17 Inverted staff

In the example shown in Fig. 5.17, the difference in height between points A and B is

$$A - B = 1.275 \text{ m} - (- 2.447 \text{ m})$$
$$= 1.275 \text{ m} + 2.447 \text{ m}$$
$$= 3.722 \text{ m which is obviously a } rise.$$

Since the R.L. of point A is 23.518 m, the R.L. of the underside of the tunnel will be

$$23.518 \text{ m} + 3.722 \text{ m} = 27.240 \text{ m}$$

If the situation is now considered in terms of the height-of-collimation method, we have

$$\text{H. of C.} = 23.518 \text{ m} + \text{reading at A}$$
$$= 23.518 \text{ m} + 1.275 \text{ m}$$
$$= 24.793 \text{ m}$$

The R.L. of the roof (point B) is therefore

$$\text{H. of C.} - B = 24.793 \text{ m} - (- 2.447 \text{ m})$$
$$= 24.793 \text{ m} + 2.447 \text{ m}$$
$$= 27.240 \text{ m (check)}$$

Once again the mathematics are quite simple, and the student should not have any problem if care is taken. The two types of level book are illustrated in Fig 5.17 to show how the principle still holds for calculating the R.L. of point C (the centre line of the tunnel).

The difference of height between points B and C is

$$B - C = (- 2.447 \text{ m}) - 2.853 \text{ m}$$
$$= - 2.447 \text{ m} - 2.853 \text{ m}$$
$$= - 5.300 \text{ m which is a } fall.$$

The R.L. of point B has been calculated as 27.240 m; therefore the R.L. of point C is given by

$$\text{R.L. of } B + (B - C) = 27.240 \text{ m} + (- 5.300 \text{ m})$$
$$= 27.240 \text{ m} - 5.300 \text{ m}$$
$$= 21.940 \text{ m}$$

The above has been calculated using the rise-and-fall method, but the student should have no difficulty in working through the height-of-collimation method to verify the result.

5.16.2 Levelling continued over an obstruction.

Where an obstruction such as a wall is encountered, it would be normal to get round it by means of *flying levels*. Although this is the best way, it may be time-consuming or so difficult as to be not worthwhile.

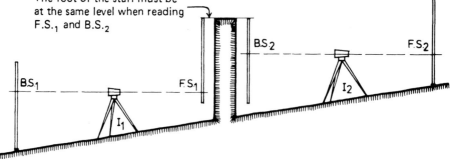

The foot of the staff must be
at the same level when reading
F.S.$_1$ and B.S.$_2$

BS$_2$ F.S$_2$

BS$_1$ F S$_1$ I$_2$

I$_1$

Fig. 5.18 Levelling over an obstruction

Figure 5.18 shows an acceptable alternative *provided that the foot of the inverted staff is held at the same height point* for both the foresight and the backsight (F.S.$_1$ and B.S.$_2$). Again the readings would be booked with a *negative* sign and the calculations would be carried out in the normal way.

5.17 Errors and adjustments

As with horizontal linear measurement, levelling is never absolutely free from error. By care and attention to detail, however, it is possible to reduce error to its absolute minimum, and experience has shown that the biggest source of error is the person using the equipment.

All forms of error must have a source or cause, and, provided that this can be determined, precautions against errors can be taken by the surveyor. Errors can be either purely instrumental or result from poor performance by the surveying team.

5.17.1 Permanent adjustments of an instrument

If instruments are faulty, good work is highly unlikely; so, to prevent a problem, it is necessary to check instruments at regular intervals (see Section 5.20). If any maladjustment is such that it cannot be corrected by the user, the instrument should be sent back to the makers. It is sometimes necessary to use an instrument which is slightly out of adjustment, but, if care is taken to ensure that the distances from the instrument to the backsight staff position and the foresight staff position are the same, the errors should cancel themselves out.

The B.S. and F.S. distances need not be measured by chain or tape, since sufficient accuracy will be attained by the use of tacheometry (see Section 5.3). Although equalising the distances of the B.S. and F.S. will prevent build-up of errors caused by inaccuracies in the instrument, curvature of the earth, and atmospheric refraction, it will not eliminate errors in individual I.S. readings, since their sighting distances will vary from the instrument station.

When levelling uphill, it will be noticed that the backsight distance will be longer than the foresight distance, while the reverse will apply to downhill

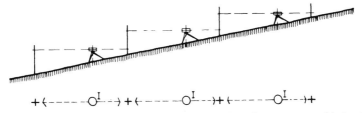

with the instrument on or nearly on line, B.S. and F.S. distances unequal in length

by means of zig-zag movement between stations, B.S. and F.S. distances can be made
equal in length, thereby reducing error

Fig. 5.19 Zig-zag working to equalise B.S. and F.S.

work. To equalise the backsight and foresight distances, this can be overcome
by working to a zig-zag pattern in plan (see Fig. 5.19).

5.17.2 Temporary adjustments of an instrument

The type of faults most likely to occur are

a) bubble not fully centred when observations are made;
b) parallax not eliminated;
c) tripod not set up on firm ground, with the result that the legs are sinking
   and causing movement of the instrument;
d) instrument moved due to the surveyor nudging it, jerking it, holding it
   with both hands when turning the tangent screw, etc., or just simply
   kicking the tripod.

The precautions are quite simple to carry out – check all the operations
carefully and watch where you put your feet, and be careful how the hands are
used.

5.17.3 Faulty staff work

The major effect of faulty staff work is to cause the readings to be too large.
This can ensue from

a) holding the staff on soft ground;
b) not holding the staff vertically.

The remedy for (a) is obvious, while ensuring that the staff is vertical can be
done in several ways:

i) Use a staff bubble, so that the staffman can check the verticality himself without signals from the surveyor.
ii) Use the vertical cross-hair to check lateral verticality of the staff.
iii) To check verticality in the direction of the line of sight, have the staffman tilt the staff *slowly* towards the instrument and then away from the instrument. The surveyor will then book the *smallest* reading at the centre cross-hair.

A further fault is caused by the staffman being improperly briefed as to where the levels are to be taken. He should be given a plan showing the positions for the staff, and he should carry out the surveyor's instructions and directions without question. Levels must be taken where they are required – not where the staffman thinks fit. The only exception to this is that, if it is impossible to place the staff where planned, the staffman may chose a suitable alternative, provided this new point can be located.

Finally, the staffman must make sure that the staff parts are fully locked into position before readings are taken.

5.17.4 Climatic effects

It is surprising how much the climate can affect the proper conduct of a survey.

a) *High winds* may make it difficult to hold the staff still or vertical. When a 4 m staff is fully extended, even a light breeze can cause problems. To combat this, stays can be fixed to the staff or a frame can be employed.
   It is not only the staff which is affected. The tripod can be shaken by a stiff breeze, and the solution is either to use a windbreak around the instrument or simply to avoid levelling in very windy conditions.
b) *The sun* can also be a problem. If it is near the horizon it may make sighting impossible, so sighting near to the sun should be avoided. In sunny weather the telescope's ray shield is a must. The heat of the sun may affect various parts of the instrument – particularly the bubble, which may expand. An umbrella of a large size will help to shade the station from direct sunlight. Heat shimmer will also make sighting difficult, and sights which 'graze the ground' should be avoided in hot weather.
c) *Rain* is equally a problem. Drops of rain on the objective lens will cause reading problems, and the author does not believe in getting expensive equipment wet. Again the large umbrella can be used to keep the station fairly dry, and the ray shield can ward off raindrops. The author has a personal preference for a large floppy hat – not so much for his own use but to place over the level between observations whenever it is extremely sunny or drizzling.

5.17.5 Curvature and refraction

Although the errors caused by curvature and refraction are neglible in ordinary levelling, don't be tempted to sight over 200 m just because there are only a few remaining readings to be taken before finishing. It is much better to move to a new position and so avoid the need to make corrections. Remember that errors of this type will be eliminated if the B.S. and F.S. distances are equalised.

5.17.6 Reading errors

There is only one person to blame – *yourself* !
  The common mistakes are

a) reading a stadia hair instead of the centre cross-hair;
b) reading the decimals correctly but omitting the metre figure;
c) reading 'up' instead of 'down', especially with an inverted staff;
d) reading in error by 0.1 m.

  There is only one solution and that is *read, book, and read again*, coupled with *care, attention, and practice*.

5.17.7 Booking errors

Who is to blame? Yourself again!
  The common mistakes are

a) entering readings in the wrong column;
b) omitting to book a reading;
c) transposing the digits between reading the staff and writing in the book;
d) omitting location information

  All these things are, you can be assured, easily done. Again there is only one answer: *read, book, and read again*, coupled with *care, attention, and practice*.

5.18 Permissible error in levelling

In the examples of level-book pages shown in Figs. 5.9 and 5.11, it was established that the checks only proved the arithmetic and not the accuracy of the levelling. Every levelling task must, therefore, be arranged in such a way that it can be checked and the amount of error be established.
  Various methods of levelling back have already been described, and the error in levelling for these methods will be as follows.

a) When levelling back to the starting B.M. or T.B.M., the total B.S.'s should equal the total F.S.'s; the total rises should equal the total falls; and the first and last R.L.'s should be the same.
    Any difference is the error in the levelling.
b) When levelling from a starting B.M. and finishing on another B.M., the difference between the total of the B.S.'s and F.S.'s, between the totals of the rises and the falls, and between the first and last R.L.'s should be the same as the difference between the B.M. levels.
    Any discrepancy is the error in the levelling.

  There are always errors in levelling, and a limit must be set for the *permissible* or allowable error for a particular task. This limit will depend on the type of work being undertaken.
  Reading error depends on the magnification and resolving powers of the telescope and on the staff marking system. However, at a distance of 80 m to

100 m it should be possible to read the staff to 0.001 m (i.e. 1 mm), which will give a probable error of ± 0.5 mm.

For careful work of an ordinary nature on reasonably flat ground, an acceptable allowance may be taken as between ± 12 K mm and ±20K mm, where K is the distance levelled in kilometres (i.e. the length of circuit, or the distance between B.M.'s). If the ground is fairly steep, the limit may be increased to ± 30K mm; while for rather more accurate work, where the readings are estimated to 1 mm and the B.S. and F.S. distances are equalised, the limit would be ± 10K mm or finer.

When the permissible error for the task has been exceeded, it is possible that the whole of the levelling operations will have to be repeated. On the other hand, if the levelling back has been done on the original outward change points, it is possible that the error can be localised to one particular section, which will save a great deal of time. Localisation of the error will not, however, be possible if the levelling back has been via the shortest route by means of flying levels.

5.19 Adjustment of the level book

Provided the closing error has been found to be within the limits of the permissible error for the task, it must be distributed uniformly through the levels.

For work of average accuracy, it is important that the error is not accumulated. There is no point in adjusting the intermediate sights, since they do not affect any other reading and hence the accuracy of the work. This is not so with backsight and foresight readings, where any error will have *an effect on every reading after it*. The adjustment is therefore made only to B.S. and F.S. entries, which has the effect of distributing the error around all the levels in an acceptable if not an exact manner.

On a small job, all the levels would be reduced and the closing error be determined. This would then be distributed by adjusting the R.L. of each change point.

Example   If the closing error is 12 mm and there are four change points, each change point will have 3 mm applied to it. This means that 3 mm at the first change point alters each R.L. after it by 3 mm; a further 3 mm at the second change point causes an alteration of 6 mm to each R.L. after it; and so on, as can be seen from Fig. 5.20.

On the larger works of average accuracy, the closing error would be divided by the total number of backsights and foresights and each backsight and foresight would be adjusted by this amount.

Example   If the closing error is 11 mm and there are five backsights and five foresights, each one will require adjusting by 1.1 mm. In practice, adjustments would not be made to fractions of a millimetre, and nine readings would be corrected by 1 mm and one reading by 2 mm.

R.L.	Remarks	Adjustments to R.L.				TOTAL		B.S./F.S.	Adjust't
RL.1						0		B.S.1	+1
R.L.2						0		F.S.1	+1
R.L.3	C.P.1	+3				3		B.S.2	+1
R.L.4		+3				3		F.S.2	+1
RL.5	C.P.2	+3	+3			6		B.S.3	+1
R.L.6		+3	+3			6		F.S.3	+1
R.L.7		+3	+3			6		B.S.4	+1
R.L.8	C.P.3	+3	+3	+3		9		F.S.4	+1
R.L.9		+3	+3	+3		9		B.S.5	+1
R.L.10		+3	+3	+3		9		F.S.5	+2
R.L.11	C.P.4	+3	+3	+3	+3	12			
R.L.12		+3	+3	+3	+3	12			

(all adjustments in mm's)   (adjustments in mm's)

Fig. 5.20 Adjustment of the level book

For tasks of higher accuracy, where the intermediates are as important as the backsights and foresights, it is normal to level every point both outwards and inwards. This means that there will be two differences in height between every successive two points levelled, and the mean of these differences will give the correct value of the rise or fall as the case may be (provided they are tolerably close). This mean difference of height will be used to calculate the R.L. of the second point from the first.

Two level books will be used – one for the outward levelling and one for the inward – and the values will need to be abstracted into a *calculation book*. The method is used only where it is absolutely necessary.

5.20 The permanent adjustments of the level

In Section 4.5, three types of instrument were considered and it was shown that each type required certain critical conditions – known as permanent adjustments – to be maintained if accurate results were to be given. Each type of level will now be considered separately again in terms of maintaining these adjustments.

5.20.1 The dumpy level

The two critical conditions are

   i) that the bubble-tube axis is perpendicular to the vertical axis;
  ii) that the line of collimation is parallel to the bubble-tube axis.

a) To check condition (i)   Set up the instrument by means of the levelling-up procedure described in Section 5.5.2. If the bubble will *traverse* when the telescope is turned through 180° in plan, there is no problem. If it moves off-centre, however, then adjustment is required.

b) To adjust for condition (i)   Note the amount by which the bubble has moved off-centre and, by means of the foot screws, move the bubble *halfway back towards the centre*. The bubble is then moved the rest of the way back to the centre by means of the adjustment screws fitted at one end of the bubble tube.

The process of levelling-up is then repeated and further adjustments are made until the bubble *will traverse*, no matter in which direction the telescope is pointed.

c) To check condition (ii)   This is carried out by what is termed the two-peg test (see Fig. 5.21). There are several ways by which this test may be carried out, but the one described below will suffice.

Place two pegs on fairly level ground, 100 m apart, and set up the instrument *exactly midway* and on the line between them at I_1. Readings are taken on a staff held in turn on pegs A and B, being noted as a_1 and b_1. Since the sighting distances are equal, any collimation error in the level will be cancelled out. The true difference of height between A and B is therefore $(b_1 - a_1)$.

The level is now set up at I_2, on line with the pegs but outside them, and as close to peg B as the short focus of the telescope will allow for reading the staff.

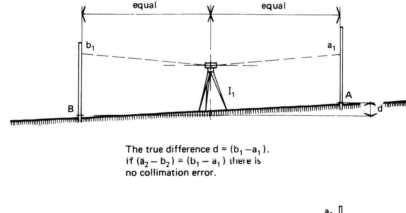

The true difference $d = (b_1 - a_1)$.
If $(a_2 - b_2) = (b_1 - a_1)$ there is
no collimation error.

Fig. 5.21 'Two-peg' test

Again the staff is read on each peg in turn, and the readings are noted as b_2 and a_2.

If $(b_1 - a_1) = (a_2 - b_2)$ there is no error. If the difference between the sets of readings varies then there is a collimation error and adjustment will be required. If the formula is written $(b_1 - a_1) = (b_2 - a_2)$ the result will be exactly the same because the mathematical 'sign' does not matter since the test is only concerned with *numerical difference*.

d) To adjust for condition (ii) Since the instrument was as close as possible to the staff placed on peg B, reading b_2 can be accepted and any error can be ignored. The true difference in height between the pegs, however, is still $(b_1 - a_1)$ and it is necessary to calculate the correct reading for a_2,

i.e. $a_2 = b_2 - (b_1 - a_1)$

The adjustment is carried out by moving the diaphragm in the telescope, using the appropriate adjusting screws, until the centre cross-hair cuts the staff held on peg A at the calculated true reading.

The process should be repeated from the beginning and further adjustments be made as, or if, necessary. The student is reminded to ensure that the bubble remains central at all times.

5.20.2 The tilting level

There is only one critical condition, which is

i) the collimation line must be parallel to the bubble-tube axis.

a) To check for condition (i) Carry out the two-peg test as described for the dumpy level (Fig. 5.21).

b) To adjust for condition (i) Calculate the correct reading a_2 required on the staff at peg A when observed from I_2. Using the *tilting screw*, bring the central cross-hair to this calculated reading, which will of course move the bubble off-centre. The bubble is now carefully recentred, *using its own adjusting screws*, and the operation is complete. Again, check and carry out further adjustments if necessary.

The student should note the very important difference between the adjustment of the two levels:

i) in the dumpy level, it is the *diaphragm* and hence the *line of collimation* which is moved;
ii) in the tilting level, it is the *bubble* which is moved, *not the diaphragm*.

5.20.3 The automatic level

Again only one critical condition is a requirement:

i) the line of collimation is horizontal when the circular bubble is centred.

a) To check for condition (i)   Carry out the two-peg test as described for the dumpy level (Fig. 5.21).

b) To adjust for condition (i)   Again calculate the correct reading a_2 and then move the line of collimation until this correct reading is given by the centre cross-hair.

The method of adjustment, however, will depend on the instrument in question. In most automatic instruments, the diaphragm is moved as for a dumpy level, without touching the compensator unit, although this certainly is by no means standard practice.

5.20.4 General comments on permanent adjustments

Only the general principles of permanent adjustment are covered by the above text – where other adjustments are required, such as for wear and tear on foot screws etc., it is better to let the manufacturer service the instrument. Remembering what was said about the importance of the instruction booklet and how it should be thoroughly read, it is nevertheless worth emphasising that a little knowledge is a dangerous thing and *an inexperienced operator who fiddles with the permanent adjustments stands every chance of making them much worse!*

Exercises on Chapter 5
1. Define the following terms: (a) a level surface, (b) datum, (c) foresight.
2. What is a T.B.M.? Why is it set up?
3. Explain *fully* by means of sketches and notes what is meant by the term 'change point'.
4. In the situation shown in Fig. 5.5, the horizontal distance between A_1 and A_2 is 15.000 m. Using the height information shown in the figure, calculate the gradient of the slope between the two points.
5. What is meant by the 'height-of-collimation' method of calculating R.L.'s? Make up a sample page of *six* bookings to illustrate the principle and effect the check.
6. Discuss the comparative merits of the rise-and-fall and height-of-collimation methods of booking.
7. a) Explain, with sketches, the levelling-up procedure for a dumpy level.
   b) Compare the above procedure with those employed when using the tilting level and the automatic level.
8. What is a diopter scale and what is its purpose?
9. How is a level moved to a new position?
10. Explain, by means of diagrams and brief notes, the difference between 'one-set-up' and 'series' levelling.
11. Define the following terms: (a) flying levels, (b) spot height, (c) reduced level.
12. Explain how you would set out a contour by (a) direct contouring and (b) indirect contouring. Discuss the comparative merits of the methods.
13. When contouring by tacheometry, the centre cross-hair reading is 1.560 m with a reduced level of 100.550 m. The constant of the instrument is 100.

a) What will be the required staff reading in order to set out the contour of 100.000 m?
b) If the lower stadia line is adjusted to read 1.600 m and the upper stadia line reads 1.750 m, what is the distance from the staff to the level?
14. What is meant by the term 'section levelling'? Use diagrams to illustrate the two types of section which can be drawn.
15. A straight sewer runs from C to D and levels were taken as follows:

B.S.	I.S.	F.S.	Remarks
1.260			B.M. 370.460 O.D.
0.110		2.435	C.P.1
	1.410		Ground level at C
1.005		2.385	C.P.2
0.880		2.815	C.P.3
	2.115		Ground level at D
		2.270	B.M. 363.820 O.D.

a) Rule out the level book, insert the above observations, and reduce the levels, applying all the usual arithmetical checks. Use both rise-and-fall and height-of-collimation methods.
b) If the distance CD is 210 m, what is the gradient of the sewer?

6

Angular measurement

It has been shown that all *linear measurements* taken for the purposes of a map or plan must be *truly horizontal*. The same is true of angular measurements. The *compass* achieves this automatically because the compass card is balanced so as to lie *horizontally on its pivot*. Ordinary levels which have a *horizontal circle of degrees* will also measure in the horizontal plane provided they have been set up so that their *temporary adjustments* are correct. The sextant, on the other hand, measures the actual angle *subtended to the observer* and, except when both objects observed lie in the same horizontal plane as the observer, this angle is a *slope angle*.

The instrument most used for the measurement of angles, whether in the horizontal or the vertical plane, is the *theodolite*. Most students look upon the instrument with horror, thinking it to be so utterly complex that to use it will be beyond them forever. In reality, the theodolite is a development of the level, in so far that it can be used in the same way but can also perform many more functions besides. The author can assure his readers that it neither growls nor bites and is certainly not an instrument of which to be frightened.

This chapter will, therefore, be an introduction to the instrument and the method of using it to measure horizontal and vertical angles. The text will not be exhaustive although it is hoped that should more advanced studies be undertaken, the instrument will not appear as forbidding as the student first thought and that it will be mastered quickly and its use enjoyed to the full. In order to grasp the basic principles and to understand the instrument thoroughly, the student is advised to read the text with a theodolite to hand.

6.1 Uses of the theodolite

When chain-survey techniques are inadequate or cannot be used, the versatility of the theodolite lends itself to the following tasks:

a) the measurement of horizontal and vertical angles,
b) setting out lines and angles,
c) levelling,
d) optical distance measurement,
e) plumbing tall buildings,
f) plumbing deep shafts,
g) geographical position-fixing from observations of the sun and stars,
h) checking lines and heights, and so on.

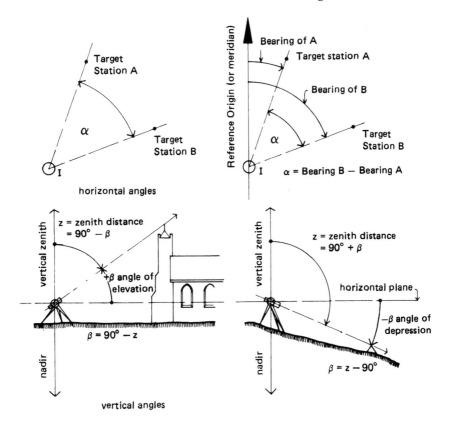

Fig. 6.1 Angular measurement

The early theodolites were big and clumsy-looking but were quite efficient in carrying out the tasks asked of them, being specifically designed for the measurement of horizontal and vertical angles in surveying and construction works (see Fig. 6.1).

6.2 Development of the theodolite

Although it is believed to have a long history, the theodolite first became well known when Thomas Digges, the son of Leonard Digges who gave the instrument its name, published a description of the instrument in 1571. Great improvements in its design were made by Jonathan Sissons at the end of the eighteenth century and by Jesse Ramsden (1735 – 1800), who is credited with inventing in 1787 the 'Great Theodolite for the Ordnance Survey of Great Britain'. This was his famous *telescopic theodolite* which had a horizontal circle 3 feet (about 900 mm) in diameter! Like all the early instruments, only horizontal angles could be measured and it wasn't until the early nineteenth century that a vertical circle was incorporated, culminating in the *transit theodolite*.

The basic concept of the theodolite is shown in Fig. 6.2, and modern instruments are refinements of the principle. Compared with the early instruments, the modern theodolite is fully enclosed so it is impossible to see how it functions. This is because the circles are made of *glass* and need to be fully protected.

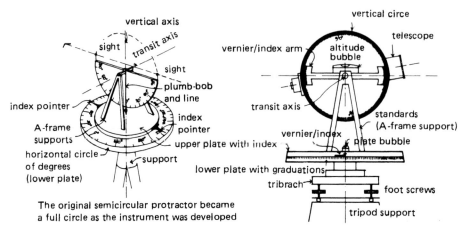

Fig. 6.2 The theodolite

6.2.1 The transit theodolite

The transit theodolite is essentially a telescope, the line of collimation of which can be *transited* (i.e. turned through 360°) in the *vertical plane*, the angles of revolution being measured on the vertical circle of degrees. The instruments can also be turned through 360° in the horizontal plane. The various parts are shown in Fig. 6.3 and consist of the following.

a) A levelling or *tribrach* system, similar to that used for an ordinary level, fits on to the head of the tripod in the normal way. The *outer tribrach* is attached to, and can be moved horizontally in relation, to the tribrach system by means of a sliding head. The tribrach is fitted with foot screws which are used in conjunction with the *upper-plate bubble-tube* to set the vertical circle vertical.

b) Two *horizontal plates* fit into, and are supported by, the *inner tribrach*. The *lower plate* (formerly known as the protractor) is marked by degrees and their subdivisions, in a clockwise direction from 0° to 360° (i.e. 0° again), while the *upper plate* carries engraved index marks (sometimes equipped with vernier scales) by which the angle of rotation is read from the markings on the lower plate. Each plate is fitted with a *locking clamp* and a *tangent screw*. The lower-plate clamp locks the lower plate to the inner tribrach, and the tangent screw allows the plate to rotate in relation to the inner tribrach by a small amount under control. The upper-plate clamp locks the upper plate to the lower plate, and the tangent screw allows the upper plate to rotate in relation to the lower plate by a small amount under control. This means that, because the upper plate's central support slots *inside* the lower plate's central support, which in turn fits *inside* the inner tribrach,

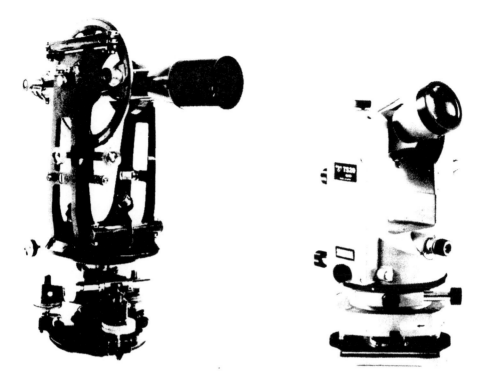

Fig. 6.3 A transit vernier theodolite (*left*) and a modern theodolite

i) when the lower plate is clamped to the inner tribrach and the upper-plate clamp is free, the upper plate and telescope can be moved independently and in relation to the horizontal circle of degrees;

ii) when the two plates are locked together, the whole instrument may be rotated about the vertical axis if the lower-plate clamp is released;

iii) when both plate clamps are locked, no movement of the instrument will occur except that

    a) if the lower-plate tangent screw is turned, both plates will *rotate together* by a small amount under control; or

    b) if the upper-plate tangent screw is turned, the upper plate alone will rotate by a small amount under control.

c) Two *A-frames* or *standards* are carried by the upper plate. These carry a *horizontal axis* known as the *trunnion axis* or *transit axis*. The trunnions of the telescope are supported by the tops of the A-frames.

d) The trunnion or transit axis has the *vertical circle* (or protractor) and the *telescope* rigidly fixed to it, and the circle and the telescope therefore *rotate together. The vertical circle is fixed vertically and at right angles to the transit axis* and is divided into 360°. Originally this was only half a circle, divided from 0° at the bottom middle to 90° at each end of the horizontal line of collimation (see Fig. 6.2). The horizontal line of collimation is defined on a full vertical circle by 0° or some multiple of 90°, and various reading systems are now available.

The movement of the circle is controlled by a *vertical-circle clamp* with a *tangent screw* for adjustment in the normal way. Optically, the telescope is the same as that used in an ordinary level.

The upper plate, including with it the standards, telescope, plate and altitude bubbles etc., is termed the *alidade*.

e) There are two bubbles:

    i) the *plate bubble*, which has already been mentioned in Section 6.2.1(a); and

    ii) the *altitude bubble*, which is fixed to the arm carrying the indexes (and verniers, if fitted) by which the angle of elevation or depression is read from the vertical circle. When this bubble is centred, the line joining the indexes is truly horizontal and therefore the vertical angle can be read directly from the circle.

f) A *plumb-bob* is hooked to the underside of the spindle which forms the vertical axis of the instrument, so that the theodolite can be centred over a specific point of ground (i.e. a survey station etc.)

g) When vernier scales are provided at the indexes, the instrument is called a *vernier transit theodolite*.

The above is the description of the working parts of a typical theodolite. The instrument shown on the left in Fig. 6.3 is over 60 years old and is, in fact, in full working order. It is a beautiful piece of craftsmanship, although its shape and size are as far removed from the modern instrument as it is possible to get. It is read by the naked eye through lenses which magnify the engraved markings on the metal circles.

Although the principle is maintained, theodolites differ greatly, chiefly in the closeness of the graduations, the size of the circle, and the devices by which these are read. Besides the verniers already mentioned, reading may be by means of micrometer microscopes or by various optical arrangements.

Optical arrangements require the graduations to be scribed on glass. This allows finer graduations to be made which, when light is reflected through the circles, are seen very brightly and clearly. The circles are also much smaller. The vernier transit theodolite illustrated in Fig. 6.3 has a vertical circle of 6 inches (150 mm) diameter and a horizontal circle of 5 inches (125 mm) diameter. Compare these sizes with the modern 'twenty-second digital theodolite', which has a vertical circle of 70 mm diameter and a horizontal circle of 80 mm diameter, all rather a far cry from Ramsden's diameter of 3 feet for the horizontal circle.

6.3 Setting up the theodolite

When setting up the instrument in the field at a station point, *the station adjustments* are three in number: (i) *centring*, (ii) *levelling-up*, and (iii) *focussing*. These can be likened to the temporary adjustments of a level.

6.3.1 Centring

When set up and ready for work, the instrument must have its vertical axis *exactly over the station mark*. The tripod is therefore set up over the mark with its head approximately in the horizontal plane, and the theodolite is carefully and securely attached to it. When lifting the theodolite, make sure that *all the clamps are on* and that is held with one hand on the standards and the other on the tribrach.

When the theodolite is secured to the tripod, *release all clamps* and centre the instrument over the mark. With older instruments this very often meant attaching a plumb-bob and line to the underside of the vertical axis, a hook being fitted to the instrument for this purpose. The legs of the tripod were then 'juggled about' until the plumb-bob, when at rest, was over the mark. When a sliding head was available, straight-line movements of the instrument enabled the plumb-bob to be centred with more speed. Most modern instruments are fitted with an optical plummet which makes centring over the mark simplicity itself. The plummet is adjusted by means of a sliding head which may be part of the tripod or part of the tribrach, but, in either case, the sliding head should be locked in position when it is not being used, to prevent any accidental or unwanted movement of the instrument.

Having centred the instrument approximately over the mark, push the legs of the tripod firmly into the ground and tighten any leg nuts. The instrument is now ready for the second stage of the setting-up operation – *levelling-up*.

6.3.2 Levelling-up

This stage of the operation is of the utmost importance and is done by means of the *foot screws and the plate bubble* in exactly the same way as described for the dumpy level (refer to Section 5.5.2). Remember that the plate bubble must remain centred when the instrument is rotated through 360°.

If the plate bubble is out of adjustment, the levelling-up can be carried out by means of the *altitude bubble*. Together with the telescope, the altitude bubble is placed parallel with two foot screws and the bubble is centred. The bubble and telescope are then turned through 180° and the bubble is again centred, *halfway by the foot screws and halfway by its own tangent screw*. It is then turned through 90° and centred *entirely by the third foot screw*. The process is repeated until the levelling is exact, following which it is advisable to adjust the plate bubble to the centre position by adjusting the screws on the level tube.

When the levelling is completed, loosen the holding bolt of the sliding head and, using the straight-line movements mentioned earlier, locate the centre of the instrument (plumb-bob or plummet) *exactly* over the station mark. *Do not rotate* the instrument in plan when sliding it on the head, as this will upset the centring of the bubble.

Finally, check the levelling-up and the centring and adjust as needed.

6.3.3 Focussing

This is very similar to the focussing of a level.

Remove the lens cap from the telescope and put it in the instrument case, or in a pocket, for safe keeping. First adjust the cross-hairs and then the main focus. Very short sights are seldom required, with the possible exception of traversing, so the telescope will be at infinite focus. It is better, therefore, to adjust the main focus on an object 700 to 800 m away from the station. If a modern instrument is being used, open up the illumination mirrors and adjust to allow maximum light into the instrument. Eliminate parallax in the usual way (Section 5.5.3). Finally, focus the reading system; i.e. if magnifiers are used unfold them to the reading position or, on a more up-to-date theodolite, focus the microscopes.

Remember that the tripod should not be touched, kicked, or brushed against once the setting-up has been completed. The student should also bear in mind that ground vibrations can affect the instrument – so do not tread heavy-footed in the vicinity of the tripod. When handling the instrument, treat it very gently and touch it only as necessary to operate it.

Finally, never leave the instrument unattended, for reasons that are obvious. It must also be protected from wind, rain, excessive sun, etc., and a surveyor's umbrella or a large floppy hat are pieces of ancillary equipment that should always be to hand. The instrument is expensive – look after it!

6.4 Moving to another station

Moving to another station is a relatively simple operation, provided certain precautions are taken to protect the instrument.

Place the protective cap on the telescope objective lens (remember, it should always be in the case or a pocket for safe keeping). All clamps should be released and the telescope should be pointed vertically. The holding bolt is then released, so that the instrument can be made central on the tripod head, following which the bolt is retightened. Centre the foot screws in the middle of their run.

Ease the feet of the tripod out of the ground and carefully bring the legs together. If a plumb-bob is being used, this should be removed before the above operation and put into the instrument case for safety. Lift the tripod and theodolite together and carry them *upright*, in front of the body, until the next station is reached. Where the terrain is rough or the stations are a long distance apart, it is perhaps better to dismantle the theodolite from the tripod and put it in its case; then there will be less risk to the instrument.

When the new station is reached, set up again as described in Section 6.3.

6.5 End of the work

The procedure is exactly the same as the above, except that at the end of the work the theodolite *must be put away correctly into its case*. Once in its case, all circle clamps should be gently applied and the holding clips should be fastened into place. Check the case to see that the plummet and other accessories are there.

Telescope and close together the legs of the tripod, fasten them together by whatever means is provided, and make sure that no loose equipment is left lying on the site.

6.6 Reading systems

Several reading systems are available, ranging from the *direct vernier* to *optical micrometers*.

6.6.1 The vernier scale

This is a device fitted to precision instruments to enable accurate readings to be made. The system was invented by a Frenchman, Pierre Vernier, at about the beginning of the seventeenth century, and consists of a small scale, *the vernier scale*, which is placed against a longer scale called the *primary or main scale*. By sliding the scales one against the other, the measurement is given.

In Fig. 6.4(a), the main scale has its major units subdivided into *ten* equal parts so that when the *index* coincides with any line on the scale it is easily

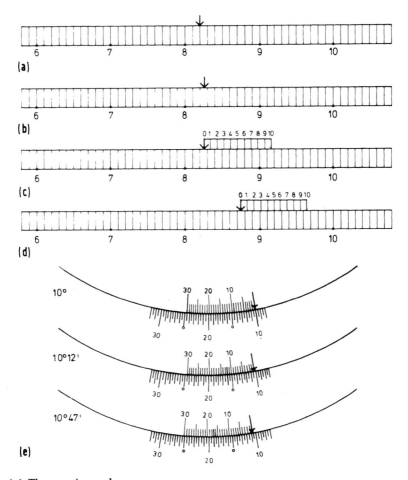

Fig. 6.4 The vernier scale

read in terms of units and tenths of a unit. The index (pointer) shown in Fig. 6.4(a) is at 8.20 units. However, as illustrated by Fig. 6.4(b), if the index lies *between* lines on the scale, the required fraction of the tenth part must be estimated, since in most cases it will be impracticable to subdivide the spaces indicating tenths into ten parts, each representing one hundredth part of a main scale unit. Using a vernier scale will enable small parts of a unit to be measured.

A vernier scale is prepared equal in length to $(n - 1)$ subdivisions of the main scale. In the examples shown in Fig. 6.4, each unit of the main scale is subdivided into *ten* parts, therefore $n = 10$. It follows then that the length of the required vernier will be equal to $(10 - 1) = 9$ subdivisions of the main scale. The vernier length itself will then be subdivided into n parts, which in the examples is 10.

This vernier is then attached to the index as shown in Fig. 6.4(c) and, by moving along the vernier scale until one of its lines coincides with a division line on the main scale, readings to a hundredth of a unit will be given. In the example, it is the sixth line of the vernier which is in line with the main scale line and the reading is therefore 8.26 units (i.e. 8.20 + 0.06).

In Fig. 6.4(d), the reading is 8.75, and the student should verify this for himself. Note also that it is normal for *the vernier scale to be read in the same direction as the main scale*.

Following the same principle, verniers found on the *transit theodolite* (Section 6.2.1) have their main scales divided into degrees and half degrees. Twenty nine of the half-degree divisions are taken for the vernier-scale length, which is then divided into *thirty* parts. One division on the vernier scale thus equals 29/30 of a half-degree division on the main scale, so the difference between them is 1/30th. This difference is 1/30 of half a degree, which is 1/60 of a degree or *one minute*. This amount is known as the *least count of the vernier*. Three examples are shown in Fig. 6.4(e), and again the student should verify the values.

6.6.2 The drum micrometer

This was introduced as an alternative to the vernier and proved to be more accurate. A movable hair-line is placed above the index in a microscope viewer and is coupled to the drum in such a way that one complete rotation of the drum moves the hair-line *laterally* by a distance equal to the width of one circle graduation.

The drum perimeter is divided into suitable parts, but when the hair-line is over the index the drum reads zero. When reading, the circle graduation *before the index* is noted, then the hair-line is moved over *that* graduation line. The parts then shown on the micrometer drum are then added to the circle graduation to give the correct reading.

Using two or sometimes four indexes, with a micrometer drum and microscope over each, micrometer instruments allow multiple readings to be taken. Although the mean reading is free of any error caused by eccentricity of graduation marking, the system is slow and tedious, so much so that conditions could change during the time taken for observations to be made.

6.6.3 Optical reading systems

Since the *glass circle* has completely replaced the brass circle as used in the vernier theodolite, optical reading systems are necessary. Theodolites fitted with these are called *optical theodolites*.

a) The circle microscope The simplest form of optical reading system (Fig. 6.5(a)) is merely a microscope to view the circle graduations against a fixed hair-line system. Circles are usually graduated to five-minute intervals, with single minutes estimated. (Note: The index hair-line will have an apparent width of one minute of arc, which may help the estimation.)

b) The optical scale One stage up from the circle microscope is the *optical scale*. A microscope is used which is fixed on the scale. Only *one degree* shows on the microscope scale (see Fig. 6.5(b)), but this is fully divided into minutes, the seconds being estimated to the nearest ten. There is no fixed index mark for this type of reading system.

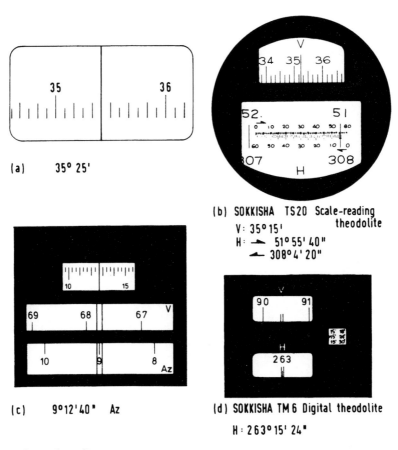

(a)   35° 25'

(b) SOKKISHA TS 20 Scale-reading theodolite
    V: 35° 15'
    H: ⟶ 51° 55' 40"
       ⟶ 308° 4' 20"

(c)   9° 12' 40"   Az

(d) SOKKISHA TM 6 Digital theodolite
    H: 263° 15' 24"

Fig. 6.5 Optical reading systems

c) The optical micrometer   This is a system whereby the optical scale and the lateral movement of the micrometer are combined. The system is very accurate and is easily read, typically to 20 seconds direct and 5 seconds by estimation. The microscope carries a central fixed hair-line or a pair of close parallel lines, and shows an image of circle graduations along with an image of a scale like the optical scale. Coupled to the system is a micrometer drum, which means that the images are displaced when the drum is rotated. The reading is therefore the circle mark at the hair-line plus the reading of the optical scale at the hair-line (see Fig. 6.5(c)).

6.6.4 Theodolite circles

The beginner is probably wondering which circle is which, and here again different manufacturers use different notations.

The *horizontal circle* is noted as either 'H' or 'Az' (short for aximuth). 'Hz' may also be used.

The *vertical circle* is noted as 'V'.

Although the examples mentioned have been with reference to circles divided into 360°, *gradiens* can also be used, circles then being divided into 400^g. This is known as the *centesimal* system.

Some manufacturers will make the two circles in different colours to eliminate mistakes, while the read-out may also be by digital means (just like a pocket-calculator read-out). It will be well worthwhile for the student to acquire literature from the various manufacturers of theodolites and study the various reading systems available – including methods of illumination for working in the dark.

6.7 Face of the theodolite and compensated measurement

Before making observations through the telescope of the transit theodolite, it is necessary to check the *face of the instrument*. This is simply the *position of the vertical circle in relation to the observer* (see Fig. 6.6).

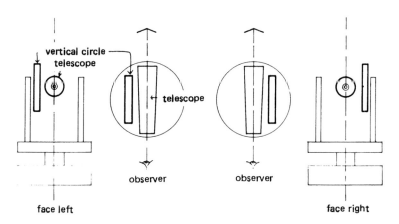

Fig. 6.6 The 'face' of the instrument

a) **Face left**   This is the normal observing position, where the *vertical circle is to the left of the observer*. The position may also be termed *circle left*. (Note: Some manuals of instruction give this as position I, in which case the letter 'I' is marked on the standard of the instrument as it faces the observer.)

b) **Face right**   The *vertical circle is to the right of the observer*. To change from face left to face right, the telescope is *transited* (i.e. reversed end for end through 180° vertically), following which the alidade is turned through 180° in the horizontal plane. The telescope will again be pointing at the target but the vertical circle will be to the right of the observer. The face-right position may be termed *circle right* or position II, as described for position I.

It is important to know the face of the instrument, as all observations are noted as either face-left or face-right values. All angles should be measured *twice* – once with face left and once with face right – and the results be *meaned* (averaged) in order to eliminate most of the small maladjustments of the instrument. One of the two values will be larger than the true angle and the other smaller, but in each case the *collimation error* will be the same amount. The *mean* of the two values will therefore give the value of the true angle. This double-observation procedure is known as *compensated measurement*.

Although it is generally good practice to measure each angle twice, the accuracy required and the quality of the theodolite being used may determine that to *measure the angle once* will be sufficient or that to *compensate and measure more than twice* will be necessary to the final result. However, the student should always bear in mind that no amount of compensation measurement will eliminate errors caused by imperfect levelling-up of the instrument.

6.8 Measurement of horizontal angles

There are several methods available, ranging from single measurement to several observations for each angle. In the following descriptions of these methods, the student should note that

a) *to turn the alidade* means to rotate the instrument about its vertical axis, in the horizontal plane;

b) *to transit* means to reverse the telescope end for end about the transit or trunnion axis and in the vertical plane.

6.8.1 Measurement of a single horizontal angle

Figure 6.7 shows the angle AOB subtended by the direction lines from station O to targets A and B. It is required to measure this angle, for which the procedures are as follows.

   i) Set up the instrument, on its tripod, exactly over the station O and ready for face-left observations.

   ii) With *all* clamps free, turn the alidade until the index of the horizontal circle is at approximately zero on the circle. Apply the *upper-plate clamp* and bring the index exactly to zero by means of the *upper-plate tangent screw*.

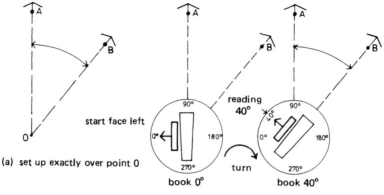

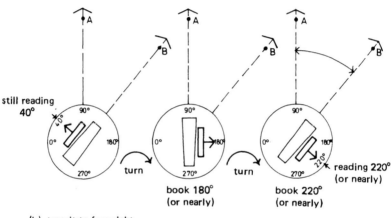

Fig. 6.7 Measurement of a single angle

(Note: Whether the reading system is a simple vernier or a circle microscope, there is no real need to bring the index coincident with zero. In *precise work*, any arbitrary circle value within the region of zero would be used as a starting point.)

iii) Turn the alidade so that the telescope is aimed on the target A and, when it is, apply the *lower-plate clamp.*

(Note: Since the alidade was turned with the *upper clamp on*, the circle reading, whether zero or an arbitrary value, will remain unchanged from stage (ii).)

Focus the telescope on the target and, by means of the *lower-plate tangent screw*, bring the central vertical cross-hair to bisect the target exactly.

iv) Check that the index is still reading 0° (if this system is being adopted). *Read and book both verniers* (or the circle microscope), noting them as *face-left* observations.

v) Release the *upper clamp* and turn the alidade so that it is aimed on the target B and, when it is, apply the *upper clamp*. Again, exactly bisect the target B with the *vertical* cross-hair but this time by means of the *upper tangent screw*.

vi) Read both verniers (or the circle microscope), noting them again as face-left observations.

The difference between the values noted when the telescope was aimed at A and then B will give the angle turned through in swinging from one direction to the other, as shown below.

Station	Target	Face left	Face right	Mean	Angle
0	A	00°00'00"	–	–	75°20'10"
	B	75°20'10"	–	–	

(Note: Only one reading of the circle is given here, for the sake of simplicity. The student should note, however, that when both verniers of a circle are read, the readings should be *meaned* to give the most accurate reading. On the vernier transit theodolite, the horizontal-circle verniers are marked A and B. There is, however, very little loss of accuracy if only one vernier is read, provided that it is the same one that is read throughout the operations.)

Where an angle only need be measured once, the above simple booking will suffice, along with the stages (i) to (vi). If, however, the angle needs to be measured with both face left and face right, the procedure must be extended as follows.

vii) Having booked the readings as at stage (vi) face left, *change the alidade to face right*, using the *upper-plate clamp*. Readings will now be read on the opposite side of the circle. Bisect the target B exactly by means of the upper-plate tangent screw, read, and book both verniers (or the microscope) in the face-right column.

viii) Turn the alidade to aim at target A, bisect it exactly, read, and book both verniers again at face-right observations. The booking will be as below (again, only one vernier reading is shown).

Station	Target	Face left	Face right	Mean	Angle
0	A	00°00'00"[1]	180°00'20"[4]	00°00'10"	75°20'10"
	B	75°20'10"[2]	255°20'30"[3]	75°20'20"	

(Note: The numbers (1), (2), (3), and (4) indicate the order in which the readings were taken. They are for the reader's guidance and would not be put in the actual field-book.)

The student should note that, because the alidade is rotated through 180° in the horizontal plane when changing face, the difference between the face-left and face-right readings should be 180°. However, due to unavoidable *collimation error*, the difference will not be exactly 180° and, as can be seen in the example above, the difference between (2) and (3) is 180°00′20″, showing the small error of 20″.

It follows then that, in order to *mean* the face-left and face-right values, 180° must first be applied to (i.e. subtracted from) the face-right value. This *new* face-right value can then be *meaned* with the appropriate face-left value. The angle required is then the *difference between the two means*.

For example, in the example above 180° is applied to face-right (4):

$$(180°00′20″) - (180°) = 00°00′20″ \text{ (new value)}$$

following which (1) and the new value are *meaned*:

$$\frac{(00°00′00″) + (00°00′20″)}{2} = 00°00′10″ \text{ (mean)}$$

Similarly, 180° is applied to face-right (3):

$$(255°20′30″) - (180°) = 75°20′30″ \text{ (new value)}$$

following which (2) and the new value are meaned:

$$\frac{(75°20′10″) + (75°20′30″)}{2} = 75°20′20″ \text{ (mean)}$$

The angle AOB is then the difference of the means:

$$(75°20′20″) - (00°00′10″) = 75°20′10″$$

In the field, only the face-left and face-right values are booked; the meaning and angle calculations are completed in the office.

6.8.2 Method of doubling

This method is used to achieve a more accurate measurement by *doubling the angle* on the horizontal circle. The simple procedure (see Fig. 6.8) is as follows:

> Set up the theodolite and carry out stages (i) to (vi) of Section 6.8.1. These will end with the telescope pointing at target B and with all the clamps applied. (Note the circle reading.)

vii) *Transit* the telescope and, *releasing the lower-plate clamp*, turn the alidade in a clockwise direction to aim the telescope on target A. When A has been found, apply the lower clamp.

viii) Bisect exactly the target A, using the *lower-plate tangent screw*.

ix) *Release the upper-plate clamp* and turn the alidade to aim the telescope on *target B again*. Apply the upper clamp.

x) Bisect exactly the target B, using the upper-plate tangent screw. Read the circle and book.

The circle reading will be twice the angle required (angle AOB), since the alidade has *swept out the angle twice*. By dividing the final circle reading by 2, the actual required angle is given, which should correspond to the very first reading noted in stage (vii). This will provide a check against gross errors.

The method may also be used as a check against reading errors, since use is made of face left and face right in combination.

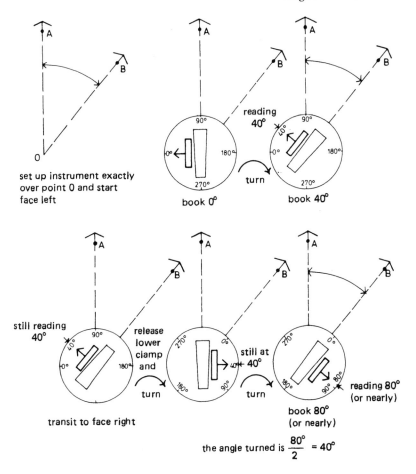

Fig. 6.8 Doubling

6.8.3 Method of repetitions

The method of doubling may be carried out several times, in which case it is termed *repetition*. For high accuracy, the angle would be swept out *n* times with face left, followed by the telescope being transited and the angle swept out *n* times with face right. Again, all turns of the alidade would be done in a *clockwise* direction.

A special *repetition theodolite* may be used for this type of work. The readings taken with face right require the alidade to be turned in the opposite direction to those with face left, which requires meaning and calculation to be done in its own special way. This, however, will be the subject of later studies.

6.8.4 Method of directions

This is an all-purpose method particularly suitable for use where several angles are to be measured from the station point. Figure 6.9 shows such a

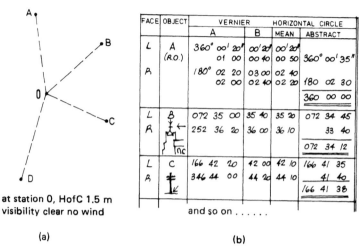

| FACE | OBJECT | VERNIER | | HORIZONTAL CIRCLE | |
		A	B	MEAN	ABSTRACT
L	A (R.O.)	360° 00' 20"	00' 20"	00' 20"	
		01 00	00 40	00 50	360° 00' 35"
R	A	180° 02 20	03 00	02 40	
		02 00	02 40	02 20	180 02 30
					360 00 00
L	B	072 35 00	35 40	35 20	072 34 45
R	B	252 36 20	36 00	36 10	33 40
					072 34 12
L	C	166 42 20	42 00	42 10	166 41 35
R	C	346 44 00	44 20	44 10	41 40
					166 41 38

at station 0, HofC 1.5 m
visibility clear no wind

(a)

and so on . . . . . .

(b)

Fig. 6.9 Method of directions

situation, where, where, from *station O*, angles relating to *the directions to stations A, B, C, and D* need to be measured.

The procedure is as follows.

i) Set up exactly over station O, as before, making all the necessary station adjustments.

ii) Using the *upper-plate clamp and tangent screw*, bring the *index of vernier A (or the microscope) coincident with zero* on the circle. Make ready for reading with *face left*.

iii) Leaving the two plates clamped together, release the *lower-plate clamp* and aim the telescope at the target at *station A*. When found, lock the lower clamp and bisect exactly the target A, using the *lower-plate tangent screw*.

(Note: Because *station A* is the first at which the telescope is to be pointed, it will be called the *referring object* (abbreviated to R.O.). All angles can then be related to the direction line from *station O* to the R.O. Because of this, the *lower-plate clamp must remain clamped* for the following two rounds of observations.)

iv) The *upper-plate clamp* is then released and the alidade is rotated through 360°, in a clockwise direction, and repointed on station A. The target is bisected exactly, using only the upper-plate clamp and tangent screw, following which the verniers (or the microscope) are read and booked.

(Note: The reason for this action is to ensure that the movements of the instrument are free, and also to obtain an unbiased reading on the R.O.)

v) Turn, aim, bisect, read, and book on stations B, C, D *and A again*, turning the alidade in a *clockwise direction* from A back to A. By reobserving A, the nearness of agreement of this reading when compared with the original will give some indication of the reliability of the observations. The technique is known as *closing the horizon* or *returning to zero*. Modern instruments should give readings which differ by only a *few seconds* of arc.

The above stages, (iv) and (v), complete *round one* of the proceedings. The *second round of angles* is carried out as follows.

vi) Leaving the bottom plate still clamped, change face to *face right* and point again at the R.O. (station A), bisecting as before using the *upper-plate clamp and tangent screw*. The reading should now differ by 180°, or thereabouts, from the original reading. Any difference, which should be small, will be due to normal collimation error. *Do not alter the setting*, but book as read.

vii) Turn, aim, bisect, read and book on stations D, C, B, *and again A*, moving in an *anticlockwise direction* from A back to A (i.e. back to zero).

The stages (vi) and (vii) complete *round two*, and for most tasks this will be sufficient. The R.O. has been *read twice with each face* to close the circle, and this procedure will check that there has been no movement in the instrument during the work.

Although changing the *direction of swing* can be considered somewhat unimportant, except in precise work, it is a good habit to have, as the technique will then come naturally when precise work is undertaken, thus ensuring good results.

If more rounds of angles are required (a further two would normally be enough, even for precise work), the *starting reading* to the R.O. should be 90°. By starting each *pair of rounds* with a different reading to A on face left, errors in graduating the circle are eliminated.

Finally, a few words are necessary in relation to the booking as shown in Fig. 6.9(b). The student will notice that by *returning to zero* at the end of each round, the number of readings to be booked for that station is doubled. However, a saving of time can be made, as mentioned earlier, by not booking vernier B readings. The student can probably see the sense of this when he remembers that with a glass-arc theodolite this is only one horizontal reading in any event. The column headed 'Abstract' is completed by *subtracting* the mean reading of the R.O. from the mean reading to the station, which in fact refers all readings to zero on the R.O. Individual angles between stations are then obtained by subtracting one direction from another.

6.8.5 General precautions

So that the angle swept out by the alidade is the correct one, it is necessary to ensure that the directions to the target are accurate. It is advisable to sight the intersection of the cross-hairs towards the bottom of the target, so that any sighting errors are kept to a minimum. For example, if the target is a simple ranging rod, it needs to be vertical. If it is not quite vertical, the necessity for it to be so will diminish the nearer to the foot of the rod the intersection of the cross-hairs is made.

Accurate bisection of the target by the vertical cross-hairs is essential and, for this reason, the target should not be so wide that the bisection is fraught with error. Similarly, the target should be clearly visible, even though presenting itself as a fine object such as the string line to a plumb-bob, a chaining arrow, etc. Care should be exercised on short sights where ranging poles can prove to be unsuitable in terms of width, and it may be better to use a proprietary survey target.

Finally, take great care to use only the clamps indicated in the described procedures. By moving the wrong clamp or tangent screw the wrong plate may be moved and a false reading be given, which will be impossible to detect until the end of the job, usually leading to a great deal of abortive work.

6.9 Measuring horizontal angles by prismatic compass

The *prismatic compass* is used by surveyors for the following purposes:
a) to read magnetic bearings;
b) to find the direction of the base line with reference to north and south, so that the plan may be *oriented*;
c) to find the direction of any other survey line with reference to north and south;
d) to fill in details after the main survey lines have been measured;
e) for making preliminary traverse surveys;
f) for making rapid exploratory surveys; and so on.

The compass shown in Fig. 6.10 consists essentially of a glass-topped metal case, varying from 65 to 150 mm in diameter. Inside the case is an aluminium ring to which the needle is attached. The ring is divided clockwise from 0° to 360° (in $\frac{1}{2}$° divisions) and rotates about a jewelled centre pivot.

The prism, from which the instrument gets its name, allows the circle to be read at the same time as the object is being sighted. Sighting is across the compass case, from a fine slit at the top of the prism holder to a sighting wire in a hinged frame. It is this wire which is seen superimposed on the compass dial, so forming the index against which the graduations are read.

When observing, the instrument must be held in a horizontal position, and for very accurate work a tripod would be used rather than the hand. The figures are engraved on the ring so as to appear as if seen by direct vision, although in fact the observer, when looking through the prism with the needle

Fig. 6.10 Prismatic compass

pointing to magnetic north, is looking at the south pole of the compass. Nevertheless, a zero reading is given through the prism, which allows a direct reading to be obtained.

Another type of prismatic compass is the *wrist compass*, which is always used in the hand, and many are filled with liquid which damps the oscillations of the circle or ring.

6.10 Bearings

The *north-seeking end* of a compass needle points towards *magnetic north*, so that the longitudinal axis of the needle lies in the magnetic north–south line, which is known as the *magnetic meridian*.

Magnetic north should not be confused with either of the following:

a) **true north**, which is the direction of the meridian through a point towards the *geographic north pole* of the earth;
b) **grid north**, which is the direction of the north–south national grid line passing through a point. The UK national grid is based on a central meridian, longitude 2° west, and, because meridians *converge*, the north–south lines of the national grid can be parallel to the *central meridian only*. It follows, then, that only when a point is on the central merdian will true north and grid north be the same. Grid north is defined on the margins of O.S. sheets.

The difference between magnetic north and true north at a point is known as the *magnetic variation* and can be termed the *declination* for the point or the area. This variation differs with place, time of day, and local conditions.

The annual amount of change at the Greenwich Observatory is between 5′ and 9′. In 1590, the variation was 11° 36′ E, while in 1663 it was considered to be zero, with the magnetised needle pointing to true north. By 1818, the declination had reached its maximum westerly value, 25° 41′, since which the movement has been eastwards. In 1962, the declination was about 7°W, and it is at present about 4°W. Not only is its value noted on the side of O.S. sheets, it may also be found in other works of reference such as *Whitaker's Almanack*.

The principle of dealing with declination and the conversion of magnetic bearings into true bearings is explained in Fig. 6.11, and the following example should make things clear.

Example   When the magnetic declination was 10°W, the magnetic bearing of a line at Manchester, England, was observed to be 175° 30′. What was the true bearing of the line?

   True bearing = 175° 30′ − 10° = 165° 30′

If the magnetic declination had been 10°E, then

   True bearing = 175° 30′ + 10° = 185° 30′

Interference from electrical cables, iron ore, etc. is known as local attraction. All reduce the accuracy and use of the compass, and in some parts of the world – especially South Africa – magnetic rocks make the use of the compass impossible.

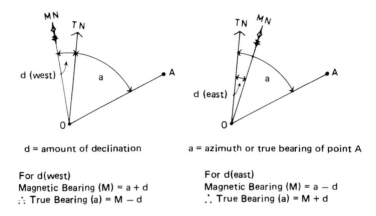

d = amount of declination

a = azimuth or true bearing of point A

For d(west)
Magnetic Bearing (M) = a + d
∴ True Bearing (a) = M − d

For d(east)
Magnetic Bearing (M) = a − d
∴ True Bearing (a) = M + d

Fig. 6.11 Declination of magnetic bearings

It is fairly obvious then that the term 'north' is a loose description, since *three* different northerly directions can be referred to on plans. It is therefore necessary to know which direction is referred to and, when *magnetic bearings* are given, the date, the declination, and the annual rate of change for the locality should also be recorded in the field notes and on the finished drawing(s).

The early mariner's compass was generally marked to give the directions N, S, E, and W. Further bisections gave a large number of named directions, known as the *points of the compass* (see Fig. 6.12(a)). Although the system is still used for navigation, surveying makes full use of the *bearing method* of describing directions. Such bearings are expressed as *degrees of arc*.

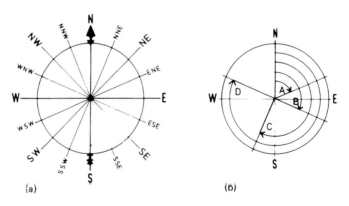

Fig. 6.12 The points of the compass and bearings

6.10.1 Bearing of a line

The *bearing of a line* is the angle which it makes with the meridian, and may be

a) *a magnetic bearing*, if it is stated with reference to the *magnetic meridian*; or

b) *a true bearing*, if it is stated with reference to the *true meridian*. The true bearing of a line which has been measured in a *clockwise* direction from true north is described as an *azimuth*. The azimuth of a line is then the angle between the line and the meridian through the observer's position. Since measurements in the horizontal plane are often referred to as *measurements in azimuth*, this is the reason for some horizontal circles being labelled 'Az' (see Section 6.6.4).

It is convenient to measure *all* bearings with reference to the north as the starting point or *reference direction*, and to make the measurements in a clockwise direction. This type of bearing is termed a *whole-circle* bearing (W.C.B.) and may have any value between 0° and 360° (i.e. north back to north).

In Fig. 6.12(b), the bearings of the lines are noted as A, B, C, and D, and some examples of bearings are given below.

N 90° E means from *reference direction north*, turn *eastwards* through an angle of 90°.

N 135° E means from reference direction north, turn eastwards through an angle of 135°.

The student should note that the *reference direction (north)* is specified first, then the angle to be turned, followed by *east*, to indicate the direction of swing. Although which north has not been stated in the above examples, this would be made clear somewhere in the field notes and could then be stated on the final drawing(s).

To find the magnetic bearing of a line, the compass is either held in the hand or is placed on a stand – remember it must be horizontal – at any point on the line. The point should be free from magnetic interference, and the observer should remember to remove keys and other metallic objects from his or her person. The compass is then turned until the sights are in line with one of the ends of the line and, when the card or needle is steady, the bearing is read through the prism.

To find the angle between two lines by means of the prismatic compass, the bearings of each line must be determined. The angle is then found by simple subtraction.

Example   In Fig. 6.13(a), the bearing of line OX is 40° and that of line OY is 300°. Find the angle between them.

The angle between them *externally* is 300° – 40° = 260°, while the internal angle is 360° – 260° = 100°.

Finally, it must be noted that for some purposes, such as traverse-survey work, *quadrant bearings* are used. Unlike the W.C.B., which has a limiting value of 360°, a quadrant bearing cannot exceed 90°, or *one quadrant*. It can be measured from north *or* south and in either case can be measured clockwise or anticlockwise. For example,

S 45° E is a quadrant bearing for SE;
W.C.B. N 280° E can be quoted as a quadrant bearing of N 80° W.

The student should note that, just as with whole-circle bearings, the starting reference direction is written first, then the angle, followed by the direction of swing.

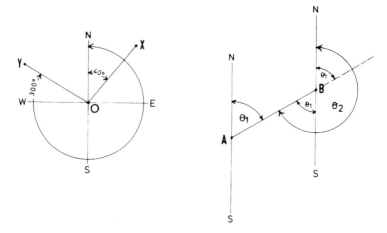

Fig. 6.13 Forward and back bearings

6.10.2 Forward and back bearings

It is obvious from the previous section that the direction of a line with reference to the *magnetic meridian* can be found by taking a bearing at *either end* of the line.

In Fig. 6.13(b), the bearing of the line joining stations A and B is the angle θ_1 when taken at *Station A* and is the angle θ_2 when taken at *station B*. It is normal to assume that the meridians through A and B are parallel, so that $\theta_1 = \theta_2 - 180°$, which is to say that the bearings of a line observed from each end should differ by 180°. If this is not so, then the error will most likely be due to local attraction, although a faulty instrument should not be ruled out.

Every line has two bearings: a *forward bearing* and a *backward bearing*. The forward bearing of the line AB (Fig. 6.13(b)), as suggested by progress *from A to B along the line*, is θ_1 or N θ_1 E and the back bearing, as suggested by sighting *from B back to A*, is θ_2 or S θ_1 W. It follows, then, that *the back bearing of AB at station A is the forward bearing of BA at station B.*

6.11 Traverse surveys

The basic principles have been mentioned briefly in Section 1.4.3 from which it can be seen that a traverse is *a sequence of straight lines the directions and lengths of which have been measured*. Where the directions are measured by theodolite, the operation would be called *a Theodolite Traverse*, whilst the use of the prismatic compass gives rise to the term *Compass Traverse*. The lengths of the lines may be measured by means of a chain, tape, tacheometry, EDM or even by *simple pacing*, depending on the accuracy required.

To enable the student to appreciate the significance of bearings, coupled with linear measurement, as the basis of truth of the statement '*a line may be completely defined and represented on a plan when its length and direction and the position of its starting point are known*', simple traverse only will be covered by this text. The rather more complex operations being left to more advanced study. However, the basic principles remain the same and, once learned, are easily extended.

6.11.1 Theodolite Traverse

a) **General fieldwork**. Four surveyors are generally required to carry out a traverse survey and their duties are:
   i) *To select suitable stations,* which is done through a reconnaissance survey of the area, looking for station positions which are fairly permanent. A masonry nail with a circle painted round it, in-situ concrete blocks with a bolt to identify the exact centre point, are just two methods of marking a station.
   Governing factors in the selection of stations include:
   (a) *Easy measuring conditions* to enable linear measurements to be taken of comparable accuracy to the angular measurements. Measurements taken through long grass, undergrowth, very undulating ground or heaps of debris should be avoided.
   (b) *Avoidance of short lines* to reduce errors from inaccurately bisected targets at short range. Sighting an instrument 2mm off target at a range of 10m equates to sighting 40mm off target at 200m range.
   (c) *Seeing the actual station mark* will substantially reduce errors. Where a ranging rod has to be erected at a station *it must be plumbed exactly above the mark* if large angular errors are to be avoided when sighting on top of the rod. (Better to sight to the *foot* of the rod whenever possible).
   (d) *The possibility of 'through' bearings* will enable a check to be made on the work (see Fig. 1.7).
   Although the factors (a) to (d) may be satisfied, it is also useful to choose stations which are near to some permanent features such as a lamp standard, corner of building, etc., in order that they may be easily found, by measurement from the feature, at a later date.
   ii) *To measure the distances between stations* in such a way that *a true plan (i.e. horizontal length of every line between stations)* can be drawn. A true plan is necessary to the calculation of rectangular co-ordinates and corrections need to be applied to directly measured lengths in order to produce true horizontal lengths. It is better in many cases to measure *slope length* and observe the *vertical angle* in order to calculate the horizontal length than to try to measure the horizontal length directly and apply corrections (see Sections 2.4.6 and 6.14).
   The equipment required is fairly basic, comprising
   • a steel band, graduated throughout in metres and centimetres, with the *zero marks* about 200 to 300mm from the ends.
   • a spring-balance calibrated in kilogrammes or a BS tension handle.
   • at least two ranging poles and several arrows.
   • *marking plates* for use on soft ground. These are usually 100mm square with the corners bent down to form spikes. (N.B. on hard surfaces, chalk the surface then carefully make the 'mark' in pencil.)
   iii) *To erect, attend and move sighting targets from station to station* as directed by the surveyor in charge of the operations.
   iv) *To measure and then record the angles* in tabular form (see Fig. 6.14).
   v) *To reference the stations for future use.*

Instrument station	Observed stations	BEARING o ′ ″	Line and length (m)	
A	B M E	47 30 00 91 00 00 128 00 00	AB	85,640
B	A M C	227 30 00 165 00 00 86 30 00	BC	83,200
C	B M D	266 30 00 226 30 00 159 30 00	CD	102,700
D	C E	339 30 00 252 00 00	DE	104,290
E	D M A	72 00 00 356 00 00 307 45 00	EA	105,450

Fig. 6.14 Booking angles

The duties listed at (iii), (iv) and (v) are self-explanatory and will become readily understood as a study is made of the procedure to be adopted when carrying out a theodolite traverse.

b) Procedure for measurement of horizontal angles. The theodolite does not measure bearings directly and in order to seek magnetic or true north, an attachment such as a *tubular compass* would have to be fitted. However, for general surveying within the construction industry, rather than fit such an attachment, it is normal practice to select a well-defined *reference object (RO)* (see Fig. 1.7) such as a tall chimney, church spire, etc., as the reference point for the survey. The bearing between the RO and the first survey station can be obtained fairly accurately from the O.S. plan and checked on site. When a survey must be oriented to true north, it is necessary to begin on two Ordnance Survey points, the bearing, length, co-ordinates and reduced levels of which can be obtained (for a fee) from an O.S. office local to the area being surveyed.

Angular measurements, also termed *angular observations*, should be taken before linear measurements so that whilst the instrument is still set up, the required vertical angles can be observed and the steel band aligned as necessary. The procedure is

i) set up the instrument at station A on face left;
ii) measure the angle between the RO and station B;
iii) obtain the face right value and calculate the mean value of the angle;
iv) repeat the foregoing stages, setting up on stations B, C, etc., and reading back to the previous station and forward to the next.

Note: As in normal levelling operations, a system of back and forward observations is used. Above, the RO is the *back station* and B is the *fore station*. For a set up on B, station A becomes the back station and C the fore station and so on.

A target must be placed at B and subsequent stations and, in the absence of proprietary equipment, a rod properly plumbed by means of a builders spirit level is adequate. In high accuracy work the measurement of each angle may be carried out several times and by different methods.

c) Procedure for measuring lines. Fig. 6.15 shows a line divided into sections at each change of gradient. The measurement of the vertical angles θ, θ_1, θ_2 and θ_3 will give the gradient of each section (with the instrument set up at alternate changes of gradient) whilst the procedure for measuring the line with accuracy is as follows.

   i) Station *two men at the back end and two men at the forward end of the steel band* to put into effect the following procedures;
  ii) Lay-off marker plates (or chalk marks, depending on the ground surface) at approximate intervals of 29.8m using a 30m tape.
 iii) Align plates (by eye or theodolite) drawing pencil crosses on them.
  iv) Lay a steel band over the first two plates.
   v) Anchor the forward end of the band by putting a ranging rod through the handle and into the ground, the rod then being held by man no. 1.
  vi) Attach a spring balance to the back end of the band and, after tightening the band, anchor the whole by putting a ranging rod, held by man no. 2, through the handle of the spring balance.
 vii) Lever back on the above rod until the correct pull of 5kg is registered on the spring balance, at which point the command 'read' is given by man no. 2 to the 3rd and 4th members of the team positioned at opposite ends of the band.
viii) The tape is read against the pencil marks and both readings, along with their difference, are entered into the field-book on line 1 (see Fig. 6.16).
  ix) Man no. 1 moves his ranging rod slightly and stages (vii) and (viii) are repeated to give a second set of readings which are entered on line 2.
   x) A third set of readings is obtained as in stage (ix) and entered on line 3.
  xi) The mean of lines 1, 2 and 3 is calculated and entered in the appropriate column as the length of the line.
 xii) Successive sections are then measured in identical manner.
xiii) The total length *AB* is then checked for gross errors against the distance measured by the 30m tape.
 xiv) The field data is used to calculate the true horizontal length of each section, and hence, the overall length of *AB*.

d) Plotting the traverse can be done either
   i) *Graphically* by laying-off the angles of the legs of the traverse, on the plan, by protractor and setting-out the lengths of the legs by scale-rule (see Section 1.4.3). This method is ideally suited to the plotting of a compass traverse because the accuracy of the plotting matches the accuracy of the fieldwork, i.e. to within 0.25 of a degree, or
  ii) *By co-ordinates* in order to take advantage of the greater accuracy achieved when measuring angles by instruments, such as the theodolite, but which would be lost if the traverse was plotted graphically.

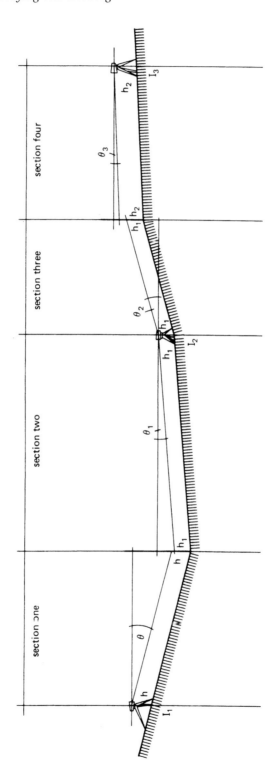

Fig. 6.15 Line divided into sections at each change of gradient

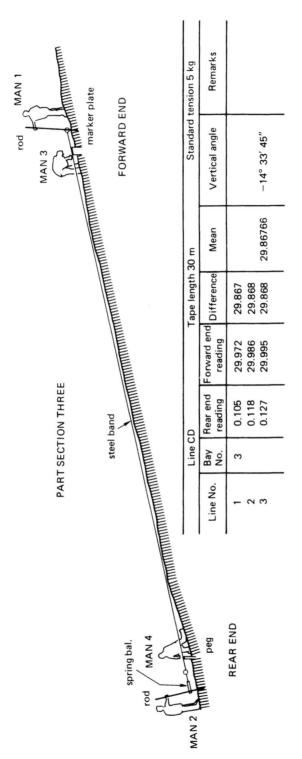

Fig. 6.16 Procedure for measuring lines of a traverse

Before plotting can be considered further it is necessary to extend the work of *offset/co-ordinates* (begun in Section 1.4.2) in relation to the traverse survey. There are three types of co-ordinates

- (a) *Polar co-ordinates* which are the basis for traverse fieldwork and compass traverse plotting, and which require no further explanation than that given in Section 1.4.3.
- (b) *Rectangular co-ordinates*, a system which is the basis of theodolite traverse plotting. In Section 1.4.2, reference is made to the locating of a point on a graph. This is because the position of a point in plan is specified by giving its *perpendicular distances from two previously fixed axes*. The axes, as on a graph, are *at right angles to each other* and intersect at *zero point*, termed the *origin of the co-ordinate system*. The easting and northing of any point are, collectively, the *co-ordinates* of the point and it is normal practice to quote the easting first then the northing. Where the whole of the survey lies to the north-east of the origin, all co-ordinate values will be positive while distances measured west or south of the origin will need to be prefaced by a −ve sign (see Fig. 6.17(a)). Note: a survey based on the National Grid System is entirely positive.
- (c) *Partial co-ordinates* which are a means of defining a line. This is shown in Fig. 6.17(b), where a line AB has been surveyed and the general descriptive terminology is indicated. In Fig. 6.18 the co-ordinates, in metres, of point A are given as +300, +400 and the length of AB as 200m. The bearing of the line AB is N30°E .

In order to plot the line AB by the co-ordinates system, it is necessary to calculate the co-ordinates of point B. This is done using basic trigonometry as follows, where

PE_{AB} is parallel to the reference latitude, and

PN_{AB} is parallel to the reference meridian.

$$\therefore \frac{PE_{AB}}{\ell} = \sin 30°$$

$$\therefore PE_{AB} = 200 \times 0.5 = 100m$$

The easting for point B, then, is the sum of the easting for point A plus the partial easting of line AB, i.e.

easting of point A   = +300
                        +100
                       ———
∴ easting of point B   = +400m
                       ———

Similarly, by calculation (which should be verified by the student as an exercise)

$$PN_{AB} = 173.2m$$

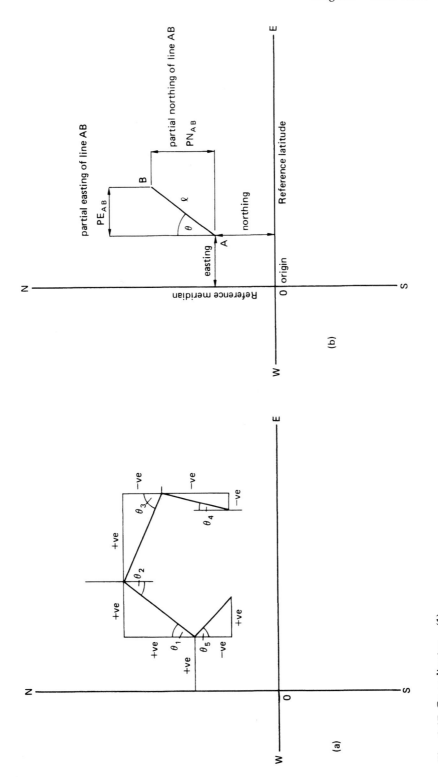

Fig. 6.17 Co-ordinates – (1)

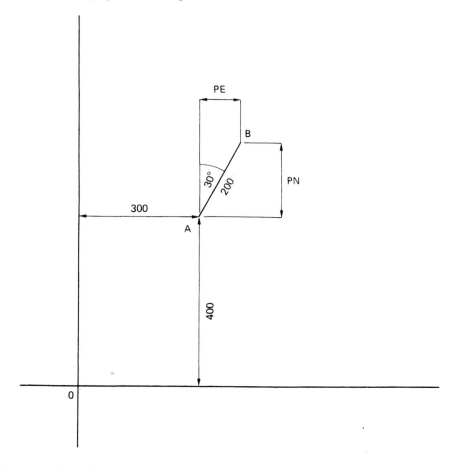

Fig. 6.18 Co-ordinates – (2)

giving the northing of B as

> northing of A = +400
> +173.2

∴ northing of B = +573.2m

From the above the following rules for calculation can be seen as

> partial easting   = line length × sine θ
> partial northing = line length × cos θ

Although the foregoing is a simple example designed to cover the basic theory of co-ordinates, the student must appreciate that *points have co-ordinates* but *lines have partial co-ordinates*. If the co-ordinates at the start of the survey are known (either real or assumed values) along with the length of each line and its bearing from the reference meridian, then it is a simple operation to calculate partial co-ordinates for each line. By successive addition of these co-ordinates to those of the start point, the co-ordinates of all the stations may be calculated. The traverse may then be plotted by drawing the axes and locating each station by scaled distances from the axes (see Fig. 1.5).

Generally, the plotting is simple, fast and, by using a drawn grid of reciprocally acting perpendicular and parallel lines in preference to the plotting of angles, extremely accurate. Although the grid spacing will vary with the scale of the drawing being produced, the actual drawn intervals will be 200mm, representing 20m at 1:100, 40m at 1:200, 100m at 1:500 and so on. Grid intersections should be marked with blue ink crosses (for permanence) with the remainder of the grid lightly drawn in pencil. Once the grid has been drawn, the procedure is:

 i) Plot each survey station from its calculated co-ordinates;
 ii) Scale-off the length of line between each pair of stations and check this against the surveyed length. Discrepancies are more often than not the result of inaccurate plotting of stations;
 iii) Check bearings by protractor for approximate value;
 iv) Draw in the lines between stations, if detail is to be plotted. The detail is plotted using off-sets and ties in the normal way.

Note: Tacheometrically obtained detail may be plotted by bearing and distance using a circular protractor (see Section 1.4.3)

In summary, traverse survey is a simple technique, the accuracy of which will vary with the type of equipment, the observational procedures used and the amount of care taken in the field.

6.11.2 Compass traverse

Although normally thought of as a means of carrying out a rapid exploratory survey, *the compass traverse* can be remarkably accurate if care is taken. It is executed in exactly the same manner as a theodolite traverse (Section 6.11.1) except that *the bearing of each line is measured from the magnetic meridian* with the result that angular errors are not cumulative.

Over the years, the compass traverse has been carried out using such instruments as *a circumferentor (surveying compass)* to the rather more usual *prismatic compass* (see Fig. 6.10) used today. The compass is generally held in the hand and the needle *floated-free* before each reading to establish the magnetic meridian relative to the actual observation being made. To prevent mistakes in reading the compass, always point the north end of the compass box towards the object whose bearing is being observed, and always read the north end of the needle. Although there is less likelihood of error in reading the modern prismatic compass, the procedure is, nevertheless, worth following.

a) **Equipment**
 i) *The prismatic compass*, which is fully described in Section 6.9.
 ii) *Measuring equipment* takes the form of *chains/tapes* for measuring length of lines between stations (*simple pacing* may suffice for exploratory surveys), off-sets and ties, and *ranging rods, arrows and pegs* for marking station points and end of chain/tape lengths.
 iii) *Noting equipment* in the form of a *survey book* in which to sketch the traverse and note down information, along with *pencils, eraser, penknife and pencil sharpener* are generally all that is required.

b) Fieldwork

Fig. 6.19(a) shows a traverse *A,B,C,D,E* in which the bearings and lengths of lines have been measured along with 'internal' detail picked up by means of offsets and ties. The procedure is as follows:

i) Choose stations *ABCD and E* so that the lines joining them are as long as possible without diverging too much from the internal detail (in order to keep offsets as short as possible (see Section 2.1.7)).

ii) Mark the station positions by inserting ranging rods into soft ground (use proprietory stand if on a hard surface). If a more permanent mark is required, or the compass is to be tripod-mounted over the mark, peg and nail etc., may be better as previously described.

iii) Take bearings as follows,

a) set up at *station A* and observe the *forward bearing to B* and *the back bearing to E*. Also observe the bearing to the feature *M* (in this case a War Memorial).

b) set up at *successive stations* and observe forward and back bearings to adjacent stations and as many bearings as are required to fix point *M*. (Bearings may also be taken to fix any other relevant points which may be inaccessible or difficult to fix by direct measurement.)

iv) Book information *up the page* as for a chain survey (Section 2.7.1) as shown on Fig. 6.19(b). *All bearings* are entered along with the length of the traverse lines.

Note: *The back bearing of each line should differ from the forward bearing by 180°.* This is proved from the geometric theorem whereby

'If a straight line cuts two parallel straight lines it makes (a) alternate angles equal and (b) the interior angles on the same side of the cutting line supplementary.'

The student should verify the difference between back and forward bearings of a line by extending the line beyond the meridians passing through the station points at each end of it and applying the above theorem.

When the difference is not 180° it suggests the presence of *local attraction* which requires correction or a mistake in the reading of one or both of the bearings.

c) Errors, corrections and adjustments

Apart from the normal errors in reading, marking-out and general lack of care, which can occur in any survey, the major effect on the accuracy of a compass survey is from *local attraction*. This affects the compass needle – and hence the reading – and has been known to result from:

i) The observer having metal jacket buttons, metal pen, metal penknife and even a steel band in his hat!

ii) The glass cover of the compass being 'electrified' by friction (discharge by touching the glass with a wet finger).

iii) Iron, steel and nickel objects near to station points.

iv) Electricity and currents from overhead lines and underground cables etc.

In town it is practically impossible to get away from local attraction such as iron pipes in the ground, lamp posts, railings, etc., and in these situations, obviously, the compass should not be used. Deposits of iron and other

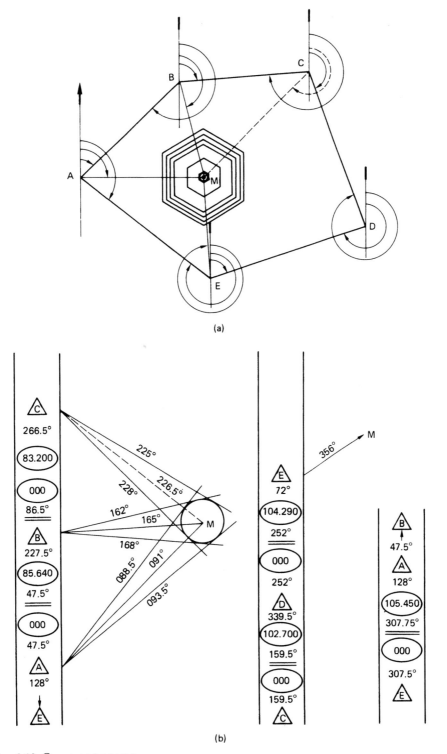

(a)

(b)

Fig. 6.19 Compass traverse

magnetic ores in the ground attract the needle and in parts of some countries, South Africa is a prime example, compass work is out of the question.

It has already been stated that the difference between the forward and back bearings of a line should be exactly 180°. For the survey shown in Fig. 6.19, the bearings are as follows:

bearings

line	m length	Stn	fwd	back	diff°	mean	Remarks
AB	85.640	A	47.5°	227.5°	180°	47.5°	Bearing M = 91°
BC	83.200	B	86.5°	266.5°	180°	86.5°	Bearing M = 165°
CD	102.700	C	159.5°	339.5°	180°	159.5°	
DE	104.290	D	252°	72°	180°	72°	
EA	105.450	E	307.5°	128°	170.5°	307.75°	Bearing M = 356°

It can be seen that the difference between forward, and back bearings at stations *ABCD* is exactly 180°. However, at station *EE* the difference is 179.5° and, because the error is no greater than one half of a degree (the accuracy to which the compass can be read), the *mean bearing* may be accepted as the true bearing of the line *EA*, that is *307.75°*.

Had the difference been greater than $\frac{1°}{2}$, say 2°, then this suggests that station *E* is being affected by local attraction and a correction of ±2° would have had to be applied to all bearings taken at station *E*, e.g.,

EA	105.450	E	306°	128°	178°	47.5°	Bearing M = 354
		(correction + 2)					+ 2
			——				——
			308°	128°	180°		356

The bearings taken to fix the War Memorial, *M* are conveniently noted in the Remarks column and, although two intersecting lines only are required, a third may be observed as a check. In the example above, because *BME* is almost a straight line, *CM* would have been a better check.

Having plotted the traverse, if the end of the line *EA* does not coincide with point *A* this is termed a *closing error*. The traverse may be adjusted using *Bowditch's method* – *the method of least distortion* which is fully dealt with in Section 9.6.3.

6.12 Extending straight lines by theodolite

Very often, in both survey work and setting-out, a straight line needs to be extended accurately, or points must be located within such a line. The theodolite is the ideal instrument for the purpose.

Where a straight line needs to be extended, it can be done by sighting in ranging poles in the normal way or by either of the following two ways using the theodolite.

6.12.1 By transiting

In Fig. 6.20(a), it is required to extend the line BA to point C. The procedure is as follows:

   i) Set up the theodolite at point A in the normal way, carrying out all adjustments.

       If the telescope is sighted on B and then transited, a point placed on line will not produce the straight line ABC but either ABC_1 or ABC_2. This is due to small unavoidable instrument errors. The correct procedure then is:

   ii) Sight on B, with face left.

  iii) Transit to face right and place a peg or marker on line at C_1.

  iv) Sight on B again with the instrument still face right.

   v) Transit (back to face left) and place a peg or marker on line at C_2.

  vi) Place a peg or marker at point C – *midway between C_1 and C_2*.

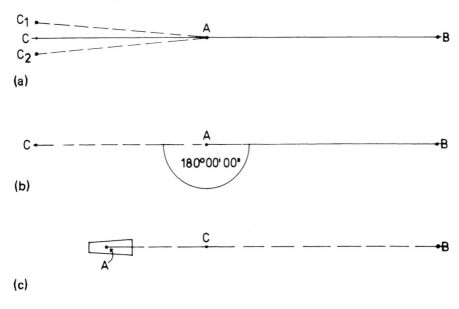

Fig. 6.20 Extending a straight line

6.12.2 By setting out an angle of 180°

The problem is illustrated in Fig. 6.20(b) and the procedure is as follows:

   i) Set up the theodolite at A as normal.

  ii) Sight on to B with the horizontal circle reading 00°00'00".

  iii) Using the *upper-plate clamp and tangent screw*, turn the alidade until the circle is reading 180°00'00".

  iv) Place a peg or marker on line at point C.

6.13 Lining-in by theodolite

Where the extremities, A and B, of a line are known and it is required to place a peg at, say, point C in the line (Fig. 20(c)), the procedure is as follows.

   i) Set up the theodolite at A as normal.
  ii) Sight on B and lock on all clamps.
 iii) Erect a pole or marker on-line at C.

6.14 Computation of true horizontal length

In Chapter 2 (Section 2.4.6) the student was introduced to simple methods of measuring horizontal distance. This was extended in Chapter 5 (Section 5.3) to include simple tacheometry using a tilting level and vertical staff. However, horizontal distance measurement may be carried out by theodolite when the sight-line is not horizontal.

When the ends of the line being measured are on different levels, requiring the telescope to be elevated or depressed (see Fig. 6.21) the normal formula

$$distance\ (D) = staff\ intercept\ (s) \times 100$$

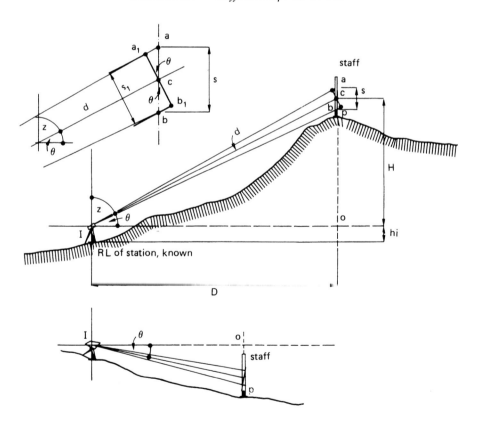

Fig. 6.21 Distance measurement with non-horizontal sight-line

needs modification because the observed intercept (s) is not at right angles to the line of sight as was the case in Fig. 5.6(b).

In order to calculate D, the *slope distance* (d) must first be calculated from the formula $d = s_1 \times 100$. Although s_1 cannot be measured it can be calculated by means of simple geometry and trigonometry. In Fig. 6.21, triangles aa_1c and bb_1c are similar and, for all practical purposes, angles aa_1c and bb_1c are right angles. In addition, from geometry, angles a_1ca, b_1cb and θ are equal. Therefore

$$\frac{a_1c}{ac} = cos\ \theta$$

$$\therefore a_1c = ac.cos\ \theta$$

similarly $b_1c = bc.cos\ \theta$

It follows then, that

$$s_1 = ac.cos\ \theta + bc.cos\ \theta = s.\ cos\ \theta \qquad (1)$$

and the following calculations can be made

a) *The calculation of the horizontal distance D*

$$d = s_1 \times 100$$

$$\therefore d = s.cos\ \theta \times 100 \qquad \qquad from\ 1\ above \qquad (2)$$

however $\quad \dfrac{D}{d} = cos\ \theta$

$$\therefore D = d.cos\ \theta$$

$$\therefore D = s.cos\ \theta.cos\ \theta \times 100 \qquad \qquad from\ 2\ above$$

$$D = s.cos^2\ \theta \times 100$$

Note: If the zenith distance angle (z) is used instead of the vertical angle θ the expression becomes

$$D = s.sin^2\ z \times 100$$

b) *The calculation of the vertical height H*

From triangle Ico, $Ic = d$, $oc = H$ and angle $clo = \theta$ therefore

$$\frac{H}{d} = sin\ \theta$$

$$\therefore H = d.sin\ \theta$$

$$\therefore H = s.cos\ \theta \times 100 \qquad \qquad from\ 2\ above$$

Note: if the zenith angle z is used then

$$H = s.sin\ z.cos\ z \times 100$$

c) The calculation of the height (reduced level) of the ground at the staff position

Although the RL of the instrument station will be known, the height of the *transit axis(hi)* of the instrument above this point must be measured.

Therefore, from Fig. 6.21

$$op + hi + pc = H + hi$$
$$\therefore \qquad op = H + hi - hi - pc$$
$$= H - pc$$

Since H can be calculated from $H = d.\sin \theta$ and pc is the *centre-hair reading on the staff*, the RL at the foot of the staff, p, is readily calculated.
Note: When the point p is higher than the instrument station then $op + hi$ is +ve and if lower −ve.

Finally, with reference to the accuracy of *vertical staff tacheometry (ordinary tacheometry)*, there are several possible sources of error.

i) *Differential refraction* due to the sight lines to the upper and lower stadia lines passing through layers of air of varying density. The staff intercept s is mostly read 'too short' so that the calculated horizontal distance D is not as long as it should be. This cannot be eliminated but is normally acceptable if all precautions are taken. As in any levelling operation, sight lines too close to the ground should be avoided.

ii) *A non-vertical staff* is a major cause of inaccuracy in *all* types of levelling work. In Tacheometry this results in the intercept s being too large or too small and, in order to avoid this error, the staffman should use a staff-bubble.

iii) *Climatic effects* which must be taken into account as in all theodolite or levelling work (see Section 5.17.4).

iv) *Instruments and equipment* which need adjustment or which are mis-treated will not give the best results. With care and regular maintenance modern equipment should not be a source of error.

v) *Lack of care* by the operatives in the manipulation of the instruments and equipment, i.e. setting-up and running constant checks of temporary adjustments, not fully extending staffs, etc., which will reduce accuracy no matter what operation is being carried out.

Because of the foregoing, it is difficult to state, with certainty, the accuracy of stadia measurement. Some authorities suggest a probable error in low level work of ±0.15m which would be permissible for plotting at 1:500 but not necessarily at 1:200, 1:100 and larger. However, modern *tacheometer theodolites* should give any *single distance measurement* to within 1 in 1000 and *heights* correct to within 0.02m (20mm). Self-reducing instruments are claimed to give an all-round accuracy of at least 1 in 1000. More advanced study will introduce the student to other checks and methods of achieving greater accuracy for all levels of work.

6.15 Measurement of vertical angles

The term vertical angle is understood to refer to *an angle of elevation or depression from the horizontal plane* when sighting to a distant target (see Fig. 6.1). As already noted, an angle of elevation is +ve and one of depression −ve and the observation of vertical angles is no more difficult than that of horizontal angles. Assuming that the instrument is already set up (see Section 6.3) the procedure is:

 i) Release *all* clamps and point the telescope at the target *with face left;*
 ii) Apply the *upper, lower* and *vertical circle clamps;*
 iii) Carefully *focus* on the target;
 iv) Using a *horizontal tangent screw* and the *vertical circle tangent screw*, bring the intersection of the cross-hairs *exactly* on to the target;
 v) If an *altitude bubble* is fitted to the instrument, *centre it by means of its setting screw before any readings are taken.* Obviously, if the instrument does not have a bubble, this stage can be omitted;
 vi) Observe the vertical angle face *left* and book;
 vii) Transit the telescope and turn the instrument through 180° horizontally. Observe the vertical angle face *right* and book;
 viii) *Mean* the two values obtained in stages (vi) and (vii) and book the new value. This has the effect of *eliminating index errors.*

Note: In normal practice for 'low order' work, a single measurement will suffice. For high level accuracy, instruments would be checked and adjusted before taking readings.

Care must be taken to 'get to know' the reading system of the instrument and a study of the 'user's manual' is a must. For example, a vertical circle can be graduated in *quadrants* (as in vernier instruments – see Section 6.6) in which case the vertical circle reading is the angle required (either +ve or −ve). Alternatively, the circle may be graduated from 0°–360°, as in optical instruments and when read face left the vertical circle reading is actually a *zenith distance angle* which must be *deducted from 90° to give the required angle* (refer to Fig. 6.21). With face right, the required angle is *the circle reading minus 270°.* Other types of reading systems allow the required angle to be read direct or will electronically compute it and display it digitally.

Exercises on Chapter 6
 1. By means of sketches and notes, describe the transit vernier theodolite.
 2. What is a vernier and how is it used?
 3. What is meant by 'centring over the mark'? How is this done by (a) plumb-bob and line and (b) optical plummet?
 4. Describe and illustrate *four* circle-reading systems.
 5. Explain, by means of diagrams and notes, what are meant by the terms 'face left' and 'face right'.
 6. What is a 'compensated measurement' when related to determining the value of an angle?
 7. Explain in detail the procedure for the measurement of a single horizontal angle.
 8. What is the difference between the method of 'doubling' and the method of 'repetitions' when measuring angles?

9. What is meant by the term 'closing the horizon'?
10. When can true north and grid north be the same?
11. How does declination affect bearings taken with a compass? What is this phenomenon?
12. Explain the following bearings and relate them to the points of the compass: (a) N 90° E, (b) N 270° E, (c) N 157½° E, (d) N 180° E, (e) S 45° E, (f) N 45° W.
13. The magnetic bearing of a line is observed to be 120°25′. What is the true bearing if the magnetic declination is 23°W?
14. Explain, by means of diagrams and notes, the difference between (a) whole-circle bearings and quadrant bearings; (b) forward and back bearings.
15. Explain how you would use a theodolite (a) to set out a straight line and (b) to extend a straight line. In each case, state possible reasons for the operation.
16. What types of survey require the observation of bearings coupled with linear measurement? Illustrate your answer.

7

Building surveys

The measurement of existing buildings is carried out for various reasons. As in plane-survey work, the amount of detail necessary and the degree of accuracy to be achieved must be determined in the light of the purpose for which the survey is required. Besides the 'work' aspect, the drawing of measured work has long been recognised as the best form of instruction in the principles of detail and design. Drawings made from one's own sketches and measurements are more valuable, from an educational point of view, than those of the same subject by others, and immense satisfaction as well as knowledge is to be gained by carrying out the whole process from first look to final drawing.

7.1 Purposes of building surveys

The main reasons for surveying existing buildings are as follows.

a) **Valuation** The requirement is to indicate the layout of the building(s) (floor by floor) and their relationship to the site. The drawing may be in single line, with room sizes and areas included. Such a survey will be used in the buying and selling of property, renting, rating, and mortgage assessment and will be backed up by additional information, in written form, relating to the state of the fabric and delapidations generally. Costs of repairs may also be included.

b) **Maintenance** The amount of detail required will be more than in (a). Special methods of construction will be noted, as well as the positions, sizes, and type of service runs. These items will be noted in drawn form to enable a building surveyor or maintenance engineer to maintain the good order of the property in his charge.

c) **Alterations and additions** Where the fabric of the building or the internal layout needs to be changed or added to, a detailed and accurate survey must be carried out. The drawings produced will form the basis upon which working drawings, specifications, and estimates of cost are prepared. An accurate survey drawing is necessary if the new work is to fit properly into or on to the old.

d) **Record drawings** Many fine buildings need to be recorded, and it is this type of survey which requires most detail and a high degree of accuracy. A 'measured drawing' is one which shows – exactly to scale in plan, elevation, and section – the building or artefact under survey. It takes no account of perspective but simply shows details and parts geometrically. The drawings are therefore similar to the ones produced by the architect from which the builder erected the structure. Figure 7.1 shows a measured drawing.

In any survey, sufficient measurements must be taken to enable the whole to be drawn out in the office. The drawing-out process is known as *plotting the survey*. It can be most inconvenient and costly if further visits to the site have to be made because important dimensions have been missed. Even allowing for the maxim 'take more than you need', it is vital that a system of measurement is adopted whereby the possibility of omissions is cut to a minimum.

While physically taking the measurements on site, always think about the plotting process. Information should be noted down in a clear and precise manner because, occasionally, there may be reasons which necessitate someone else drawing out the survey and the information must be intelligible to a draughtsman who was not involved in the measurement.

Where the subject is of sufficient size or importance, it is not unknown for draughting equipment to be taken on to the site to enable plotting to be carried out almost as soon as the information is collected. This also enables any measurements that have been missed to be taken, and so avoids a return visit.

7.2 Equipment

7.2.1 Noting equipment

6 mm plywood or hardboard board (375 mm × 250 mm) with bulldog clip.
A4-size pad of *plain* white paper.
Pencils – grades HB, H, F. Eraser. Penknife.

Note: The board may be larger to enable A3 or A2 paper to be used, if preferred, but avoid the use of 'squared' paper since this can confuse the beginner.

7.2.2 Measuring equipment

30 m fibreglass tape (for interior use)
30 m steel tape (for exterior use)
2 m or 4 m spring-steel pocket rule
2 m boxwood folding rule

7.2.3 Ancillary equipment

Long line and plumb-bob.
Five or six ranging poles. Two stands. One pair manhole keys. Tracer dyes. Hammer and cold chisels. Duster (for cleaning tapes, rubbing out chalk marks). Waterproof crayon/chalk. Torch. Ladder (for elevation measurements). Pair of steps (for internal use).

Note: When surveying the site of the building, additional equipment such as wooden pegs, nails, and studs for station marking will be useful. If the grounds are overgrown a machete, a billhook, or even an axe may be required.

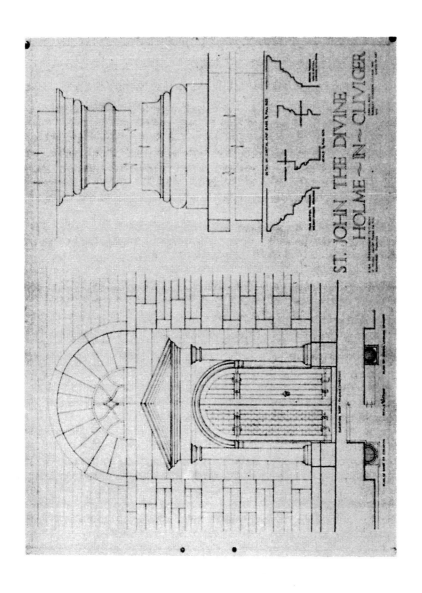

Fig. 7.1 A measured drawing

7.2.4 Plotting equipment

Board and T-square (or draughting machine). Set squares: 30° and 45° (or adjustable). Beam compass. Scale rules. Drawing instruments (compasses, dividers, etc.). Good-quality cartridge paper (preferably linen-backed) or hand-made paper. Pencils – grades 3H, 2H, H, HB, F. Waterproof inks – sepia, cobalt blue, and black.

7.3 Types of measurement

Before work can be satisfactorily measured, or the conduct of the survey be considered, the underlying principles and methods must be thoroughly understood. It is also very important to know how measurements are noted down, to enable the plotting process to be carried out successfully.

There are several ways of noting down measurements on the sketch drawings. It is necessary to bear in mind that the plotting process usually follows very closely the order in which the measurements are taken.

a) Separates – where a single measurement from point A to point B is noted. Such dimensions are usually 'ties', 'diagonals', 'thicknesses', 'heights', or 'set-backs' etc.

b) Separates and overall – where several single but consecutive dimensions are taken and noted, together with the overall dimension as a check, e.g. A to B, B to C, C to D, D to E, and A to E.

The overall dimension is necessary on two counts. Firstly, if one of the 'separates' is noted down incorrectly, the overall check will allow the error to be found. Secondly, the overall dimension will enable the standard of accuracy to be maintained. For example, where measurements are being taken to the nearest 10 mm, each 'separate' may be in error by ± 5 mm. If all the 'separate' measurements are added together, the total length will be in error by ± 5 mm multiplied by the total number of 'separates' making up the addition. To prevent this and ensure that the overall length stays within the required accuracy of ± 5 mm, the overall measurement is taken to the nearest 10 mm. The 'separates' are then adjusted to fit into the overall measurement.

c) Running measurements – where, starting at point A reading 'zero', measurements are taken in continuous form similar to chain-line technique. This system is by far the best, since, in any run of measurements, the error can only be within the reading tolerance, e.g. where reading to the nearest 10 mm the error is ± 5 mm. A further advantage is that all the measurement points within the run are automatically adjusted within the overall measurement.

Note in Fig. 7.2 how the figures of the measurements are noted down. By not using the decimal point, confusion is avoided by writing whole metre figures over millimetre figures. This convention is equally successful where imperial measurement is used.

Particular attention must also be paid to the use of dimension lines. These must not conflict with, or be confused with, the detail lines of the structure or

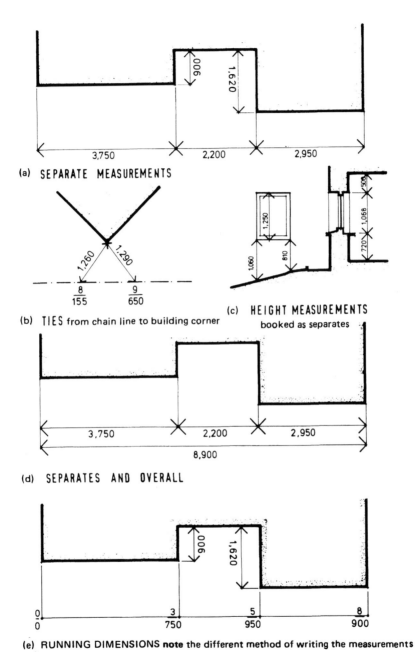

(a) SEPARATE MEASUREMENTS

(b) TIES from chain line to building corner

(c) HEIGHT MEASUREMENTS
booked as separates

(d) SEPARATES AND OVERALL

(e) RUNNING DIMENSIONS **note** the different method of writing the measurements

Fig. 7.2 Types of measurement

artefact under survey. There are also different ways of indicating the start and finish limits of a dimension and, whichever convention is chosen, this should be adhered to throughout the measuring operation. The use of arrowheads, blobs, and oblique lines are all techniques to be found on working drawings and all can be successful if used wisely.

7.4 Measuring particular features

7.4.1 Measuring simple openings

Figure 7.3 shows three simple window openings which are to be measured. The one shown in Fig. 7.3(a) will present little difficulty even to the beginner, since only two measurements need to be taken to draw the opening.

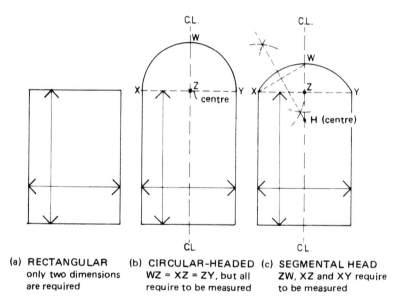

(a) RECTANGULAR only two dimensions are required	(b) CIRCULAR-HEADED WZ = XZ = ZY, but all require to be measured	(c) SEGMENTAL HEAD ZW, XZ and XY require to be measured

Fig. 7.3 How to measure different types of opening

The circular-headed window shown in Fig. 7.3(b) is not quite so simple. The width and the height to the springing-line XY pose no problems. The difficulty is to measure the curve, the height of which must first be obtained. Lay a long straight-edge across the springing-line XY and measure from this line at Z to the crown W of the arch. In the case of a semicircular arch, the measurement would be exactly one half of the total width of the opening. Using a pair of spring-bow compasses, plotting the curve to scale will be a simple operation.

However, in Fig. 7.3(c) a more difficult problem is shown, for the opening has a segmental head. The width and height to the springing-line XY are measured as before, as is the rise ZW of the arch. Provided that the curve is regular, an elementary knowledge of geometry will enable the plotting to be carried out without difficulty (this is indicated in the figure by broken lines).

7.4.2 Measuring door and window plans

Figure 7.4(a) depicts a plan through a door opening, showing how the dimensions should be taken to enable a correct drawing to be made. In taking the measurements, great care should be exercised to obtain the overall dimensions; that is, the dimensions should be taken right across the opening. Dimensions within the thickness of the opening should also be taken overall. The wall thickness and rebate dimensions are easily measured by the spring rule or folding rule. If necessary, on extra-deep reveals, a straight-edge can be placed along the wall face and measurements can be taken from this datum.

Very often, as can be seen from Fig. 7.4(b), the basic opening is measured by means of a running measurement and a larger-scale plan is made to pick up additional detail.

When plotting openings, it is very often necessary to use a centre line ('C.L.' in Fig. 7.4), to set the proper relationship between the spaces on each side of the wall containing the opening. This is particularly relevant when attempting to plot the overall internal dimension of a building alongside the overall external dimension, so that not only are the openings in their correct position but also correct return wall thicknesses are achieved (see Fig. 7.5).

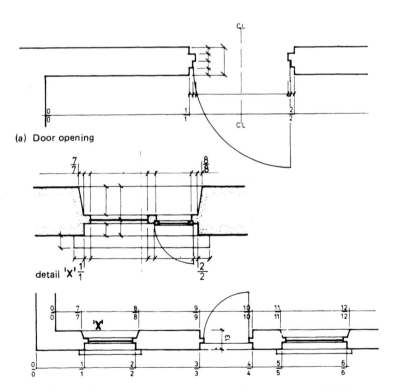

(a) Door opening

detail 'x'

(b) Small-scale notes using running dimensions can be augmented
by means of larger-scale notes of details which are fully measured

Fig. 7.4 How to note measurements of openings in plan

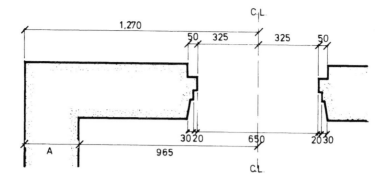

Fig. 7.5 Relating inside to outside using a centre line. By working to the centre line (C.L.) of the opening, the wall thickness *A* is automatically given when the site measurements are plotted.

7.4.3 Measuring plans of buildings

In both the measuring and the plotting of building plans, problems and difficulties will constantly arise which can, however, in all cases be solved by the application of the principle of *triangulation* or perhaps more properly *trilateration*.

It is known from geometry that, if the lengths of two sides of a triangle are known, the exact angle which they contain can be determined by measuring the third side and plotting the triangle on paper. This system of trilateration is applied to the measuring of rooms or plans, for, if the lengths of two adjacent sides of a room are known, the angle at which the two sides meet can be accurately determined by measuring the diagonal across the room or across from two points in two adjacent sides of a room, thus dividing the room into a series of triangles. It is quite rare for a room to be exactly square, i.e. having each of its four corners a right angle. Although the beginner may be forgiven for entering a room and expecting to take only four dimensions – that is, the length of each wall – it is, however, necessary to take at least five measurements to make the plan of the room complete.

Figures 7.6(a) and (b) show two rooms having exactly the same dimensions but the plans of the two rooms are very unlike. To be certain that what is drawn to scale is representative of what has been measured, the diagonals AC or BD (or both) need to be measured.

Figures 7.6(c) and (d) show other instances where complete measurement would be impossible without the use of measurements in triangulated form.

In Fig. 7.6(c) the room has one wall which is not a straight line. Only one true diagonal can be measured; therefore it is necessary to tie in the remaining room corners by a series of diagonal measurements which will enable the room to be plotted to scale. The room would be measured as follows: first measure the length of all the walls – AB, BC, CD, DE, EF, and FA – then carefully take all the diagonal measurements AC, BF, BD, and CE. Note: BF is measured before BD as it is the smaller of the two measurements and it is easier to handle the tape by lengthening it as opposed to reeling it in, which would be necessary had the larger dimension been taken first.

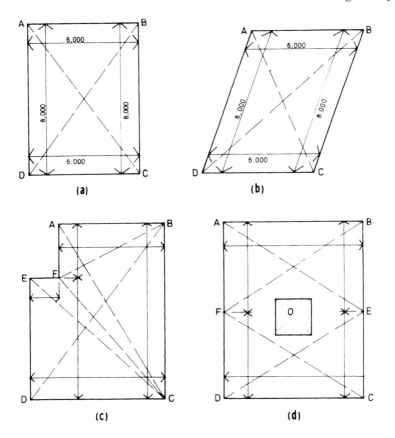

Fig. 7.6 Need for diagonal measurements and ties

In Fig. 7.6(d) the room has an obstruction 'O' in its centre, e.g. a column or a large piece of furniture, which makes it impossible to measure the true diagonals. By resorting to such schemes as the one illustrated, sufficient dimensions can be taken to enable the plotting to be accomplished.

Cases might be multiplied indefinitely in which this principle of triangulation would save both time and effort. The examples shown, however, establish the system and how it can be adapted to suit the problem in prospect.

Not all buildings requiring to be surveyed will be empty. Very often a house which is occupied needs to be measured, and great care must be taken to ensure that the occupants are disturbed as little as possible and that no damage occurs.

Measurements are usually taken at the comfortable height of one metre or about elbow height. This is just above normal window-sill height and ensures that the horizontal measuring plane shows window and other openings. Although this height is a comfortable working height which clears most obstacles such as tables and dressers, it is ideal for the tape to slice the cream off bowls of strawberries set for tea – as the author once found, to his embarrassment, as a diagonal measurement was taken.

It can be quite a costly exercise if valuable ornaments are broken due to wanton movements of the tape.

In vacant premises, it is normally quite easy to move from position to position, but movement should be kept to a minimum where premises are in use.

Referring back to Fig. 7.6(a), the room would be measured in the following sequence: AB, AC (diagonal), BC, BD (diagonal), and DA. The operation of measuring is two-handed when using a fibreglass tape. The surveyor will hold the end of the tape at the starting point and he will note down the measurements taken and called out by an assistant. (It is rare to have two assistants to carry out the measuring, leaving the surveyor free to make sketches and note down information, although this would be ideal and is to be preferred on a large survey.) By taking measurements from A to B and then the diagonal from A to C, the movement of the surveyor is minimised. Following the measurement of AC, the surveyor moves to B, while his assistant loops the spare tape over his little finger and moves to the next position of measurement from B. In other words, as many measurements as possible are taken with the surveyor in one position.

Although the above is an acceptable system, it will obviously require adaptation to suit differing circumstances. The one thing that will not change, however, is the dialogue required in taking and noting down measurements. Before starting, the surveyor will tell his assistant the positions at which measurements are to be taken. When a measurement is taken, the assistant will call out the point of measurement, followed by the measurement, e.g. 'window jamb – three metres, four twenty'. The surveyor will confirm the measurement before telling his assistant to move to the next position.

Finally, remember to tie together adjacent spaces or rooms by taking long measurements through doorways so that, when plotting, there is always a check on the accurate relationship of one room or space with its neighbour.

7.4.4 Measuring projecting features

This is relatively straightforward and should pose little or no problem provided care is taken once the principles are understood.

Figure 7.7. shows three different bay windows in plan: (a) is a square bay, (b) is a bay with splayed sides, while (c) is a segmental bay. Each will be considered separately, as follows.

a) **Squared bay**   First make a pencil sketch as nearly in correct proportions as possible. The measurements will be placed on this as they are taken.

The measurements to be taken are simple, and all that are necessary are shown in the figure. By measuring both inside and outside, and plotting equally about the centre line C.L., the return thicknesses are automatically arrived at on the final drawing. Check on site that both halves of the bay are identical.

b) **Bay with splayed sides**   Make a sketch to receive the dimensions as before. Although all the sides may be measured, these dimensions will not give the angle of the splays. This may be obtained by producing the line DC by means of a straight-edge and taking measurements as shown from A to B through to G. Care must be taken to ensure that AG is at right angles to DG. Measure GC and the splay will plot quite readily.

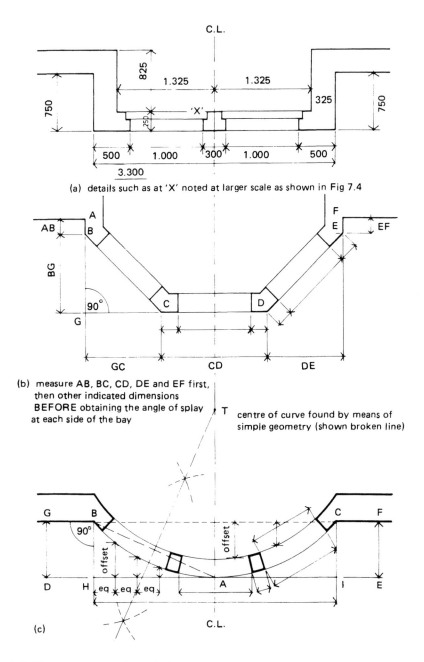

(a) details such as at 'X' noted at larger scale as shown in Fig 7.4

(b) measure AB, BC, CD, DE and EF first, then other indicated dimensions BEFORE obtaining the angle of splay at each side of the bay

centre of curve found by means of simple geometry (shown broken line)

(c)

Fig. 7.7 How to measure bay windows

c) **Segmental bay**   This is the most difficult problem of the three. Again, the first thing to do is to draw the pencil sketch, after which a base line, DE, must be laid down on the ground, if possible, either in chalk or by means of a long lath or piece of timber. This base line must be tangential to the curve of the bay at A and parallel to the main wall line at GF. Set up lines at right angles to GF at points B and C, which in turn cut DE at H and I. Take measurements as offsets (at right angles to DE) from DE to the curved face of the bay at regular intervals between H and A and between A and I.

If the curve is regular, it can be drawn using compasses, if the centre point is found by normal geometric means.

7.4.5 Measuring wall thicknesses

Very often a knowledge of construction will indicate what thickness a wall is most likely to be, but this will need to be checked. On the ground floor, doorways and windows allow this to be done relatively easily. However, problems can be encountered at first and higher floors, especially if windows will not open.

A simple method is to measure from the internal face of the wall to the glass and also from the outside face of the wall to the glass. By adding these two measurements together, plus the glass thickness (assumed as 3 mm or ignored for small scale), the thickness will be given (see Fig. 7.8).

For internal-wall thicknesses, measurements between architraves will give an indication of wall thickness – but remember that this will include finishes such as plaster.

Very often, in older property, external measurements do not tie in with internal measurements and the wall thickness is much more than expected. In such circumstances, and after dimensions have been checked, other reasons must be considered. False panelling, hidden flues, even secret chambers, are all possibilities and must be thoroughly investigated if the truth is to be drawn.

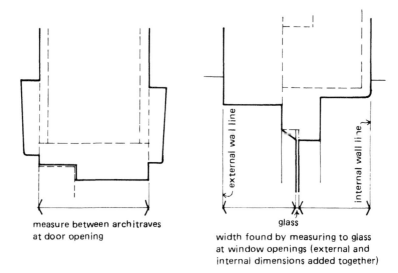

measure between architraves
at door opening

width found by measuring to glass
at window openings (external and
internal dimensions added together)

Fig. 7.8 Measurement of wall thicknesses

7.4.6 Section and height measurement

No building survey would be complete without measured sections to indicate the heights of rooms and the thicknesses of floors and ceilings. The section also gives the vertical relationship between spaces and how movement is made by means of staircases or lifts from one space to another, in much the same way as plans deal with horizontal relationships and movement.

a) **Room heights** These are measured by the folding boxwood rule once the room plan has been measured and before leaving the room.

b) **Window sill and head heights** These are measured along with room heights.

In both cases, these heights are noted on the plan sketches, being put inside circles. They may also be marked on the section sketches if this is convenient (see Figs. 7.9 and 7.10).

c) **Floor-to-floor heights** These are normally taken at staircase positions. The tape is dropped down the well from the upper floor to the lower floor (sometimes using the plumb-bob as a guide) and the measurement is noted. By subtracting the ceiling height from the floor-to-floor measurement, the thickness of the overall floor/ceiling construction is obtained.

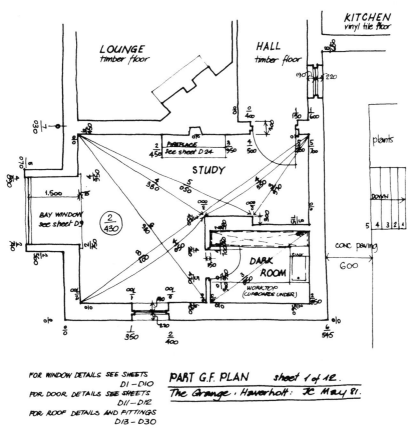

Fig. 7.9 Plan-measurement notes including height measurement

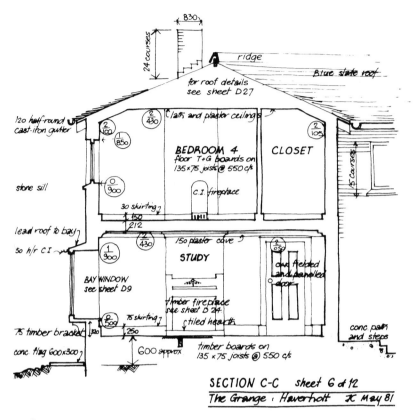

Fig. 7.10 Section-measurement notes

    This measurement will include the upper-floor finish, construction, and lower-floor ceiling finish, e.g. floor boards, joists, and plasterboard and skim.

d) Roof construction   Once the uppermost ceiling has been reached, it will be necessary to enter the loft space to determine the construction of the roof. Measurements will be taken of the ceiling joists, roof members – purlins and rafters – trusses and their centres, heights, chimney breasts above ceiling levels, and so on. All this information is necessary not only to draw the section but also to assist in the completion of the drawn elevations. It is most important that sufficient information is picked up to enable the roof slope to be drawn.

7.4.7 Elevation measurement

Measurement to enable the plotting of the elevations of the building is very often far from easy, so much so that photogrammetry is being employed more and more for complex work. This is a method whereby photographs are used in place of measurement. Obviously some linear measurements need to be known for *control purposes* but, primarily, the photograph(s) provide the information. There are two principal applications in which photogrammetry can provide invaluable information for the architect (see Section 10.5):

a) in the preparation of line drawings of extremely complex elevations and sections – but this requires special stereophotographic equipment;

b) for small projects, using simpler equipment, to give 'rectified elevation photographs' in which the inherent errors of the normal 'facade shot' are reduced or eliminated. The photographs are put into special plotting machines which, although hand-operated, produce the drawing mechanically.

However, the development of this science is far beyond the scope of the average building survey, and traditional methods still retain their place.

For all but the simplest of buildings, a ladder is required. Where a ladder is unavailable, various schemes have to be resorted to in order to obtain the necessary information.

In all cases a datum should be fixed at some recognisable level such as the top of a plinth, the damp-proof course, etc. Measurements may then be taken vertically up or down from this line. The most useful tool for this aspect of the work is the folding boxwood rule. The fibreglass tape may also be dropped from windows, and high detail such as eaves projection etc. can also be obtained from an open window.

Where the total height is required, courses may be counted and an average measurement over four courses be taken in order to calculate the total measurement over the total number of courses counted. This can be seen in Figs. 7.10 and 7.11.

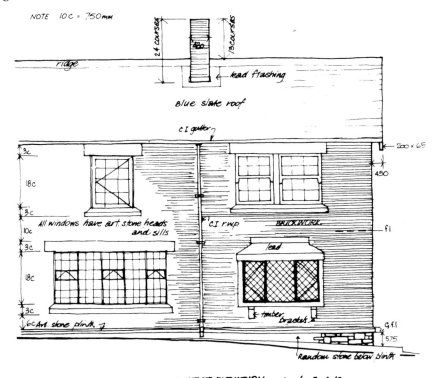

Fig. 7.11 Elevation survey notes

Provided a reasonable sketch of each elevation is done, the measurement will not prove too arduous, since all problems can be solved with a little ingenuity.

7.5 Additional information required

Once measurement is completed, all kinds of additional information must be picked up and added to the sketches.

a) **Plan sketches** *must include* types of doors and windows, including opening directions and swing; number of steps and direction of going; hatches; built-in cupboards and shelves; fireplaces and hearths (especially if, say, by Adam) and their materials; chimney breasts; changes in wall thickness; cookers; ranges; boilers; cylinders; cisterns; tanks; radiators; sanitary details including fittings and pipe runs; drainage lines, gullies, traps, manholes, downpipe positions, waste outlets.

   Staircases should include materials, direction of rise, number of risers, handrails, and general construction.

   Structural details include columns and overhead beams; direction of run of floor joists and their centres.

   Roof plans include gutters; direction of falls to flat roofs; covering materials; parapets; skylights; chimney stacks; rain-water goods (heads, gargoyles, etc.).

   Service entries must be shown – water, gas, electricity, Telecom – meter positions, socket outlets, switches, stopcocks, etc.

b) **Section sketches** *must include* storey heights; wall thicknesses; floor construction details; stair construction (if in section); roof construction; special constructions; special materials; window construction; heights of doors and lintels.

c) **Elevation sketches** *must include* horizontal dimensions where these are not shown on plan; the measuring datum; materials; rain-water goods; courses counted and average measurement over four courses; any special features.

While the above lists are far from exhaustive, they do give some indication of the information required over and above the task of taking measurements.

For example, the approximate position, or direction of flow, of a drainage line may be traced by the use of dyes. Usually a bright 'sickly' green in colour, they are put into the system at gully or manhole positions and sent on their way by running water into the drain. After a time, known outlets and other manholes on the probable line of drain are inspected for traces of the dye. It is not uncommon for coloured water to appear on someone else's land – a sure indication that the dye has not followed the path expected of it – or not to appear at all, suggesting a total blockage of the run or, perhaps, an unknown outlet, both of which require an exhaustive search to be made in the hope of finding their positions.

Figure 7.11 shows the type of sketch sheet needed to be produced in order to enable a workmanlike survey drawing to be prepared. All this information can be backed up by photographic evidence where parts of the building are inaccessible for measuring purposes or where there is fine intricate detailing such as carved cornices etc.

All sketches should be numbered within a total number of sheets; e.g. 1 of 20, 2 of 20, 3 of 20, etc. up to 20 of 20. Merely numbering 1, 2, 3, . . ., 20 is of little use because it will not be known whether or not one is in possession of a full set of information sketches. Details sketched to a larger scale on separate sheets should be cross-referenced to the master sketches of which they form a part. Finally, all sheets should contain the address of the site, the date(s) on which the survey was carried out, and the names of the persons who actually did the work.

7.6 Conduct of a survey

Although the purpose of the survey will vary, the general conduct and procedure for carrying out the work remain fairly constant. For a building in its own grounds, there is no hard and fast rule governing which area – the site or the inside of the building – should be completed first. Although it is often preferable to complete the outside and the tying-in of the building first, the weather will probably arrange the sequence without any prompting by the surveyor.

7.6.1 The site

The measurement of the site surrounding a building is carried out in exactly the same way as a typical chain survey, except that only rarely will a chain be used. For normal gardens the steel tape will be quite adequate, while for even smaller areas the fibreglass tape will be of sufficient accuracy. However, the following general procedure should be followed.

i) *Carry out a thorough reconnaissance* of the area in order to become familiar with it, noting any difficult parts to enable plans to be made accordingly.

ii) *Sketch on plain white paper* (preferably on a single sheet) the whole of the area. The sketch, while not to scale, should be of *reasonable proportions* in order that a 'feel' for the site is maintained.

Detail lines should indicate the outlines of buildings, edges of paved or grassed areas, shrubbed areas, limits of footpaths, roads adjacent to the site, walls, hedges, fences, positions of trees (inclusive of spread), telegraph poles and overhead lines, manhole covers, etc.

With an intelligent use of differing grades of pencil and a good lining-in technique, the sketch will not be as confusing as it sounds.

Parts may be enlarged in order to show more detail, provided also that reference points are used to relate the additional sheets to the master sheet.

iii) *Study the site* carefully and decide how to break the area into triangles (remember the need for a long base line upon which to hang the frame). Create check lines for the triangulations as necessary.

Certain features of the site such as walls or straight fences may in themselves be capable of acting as frame lines, and this should be taken into account wherever possible.

iv) *Finally, take all necessary measurements* and note these on the sketch sheet(s).

After all the dimensions have been noted, add any written information in the form of notes – e.g. brick paving; 900 mm × 600 mm p/c/ conc. flags; species of tree/shrub, including height, girth, and spread; type of manhole cover; and so on until the picture is complete.

7.6.2 The building

Whether or not the building is surrounded by a site, the following procedure should be followed.

i) *Obtain keys* if the building is vacant, or arrange a time of entry with the occupier. In either case, all permissions for entry on to the premises, and the purposes for doing so, must be arranged with the owner of the property. This is best done by an exchange of letters. If permission is granted in written form, this can be shown to anyone who questions your right to be in the property and this can save a lot of time.

ii) *Carry out a thorough reconnaissance* of the building. Check that all parts are accessible and that there are no dangerous areas. Is light available?

iii) *Sketch on plain white paper*

   a) a plan of each floor (including any cellars);
   b) a roof plan where there are dormer windows, ornate chimney stacks, or any special features or constructions;
   c) all elevations to the building, including any short returns which will most likely be plotted in conjunction with a section;
   d) one or more sections through the building as required to serve the purpose of the survey;
   e) any special features.

   All sketches must be of *reasonable proportions*.

iv) *Take all measurements*, noting these directly on to the sheet(s).

v) *Annotate the sheets* with additional information as discussed in Section 7.5

vi) *Before leaving* the building, check over the notes and details to ensure that there are no omissions.

vii) *Where premises are vacant*, check that all doors, windows, and gates are secured. Return keys to the owner immediately and, even though he may have been thanked orally, it is courtesy to thank him again by letter.

viii) *Where premises are occupied*, ensure that everything is as it was when the survey was started. If the occupier is present, express thanks for his or her help and also send a letter.

7.7 Sources of information

There are sources of information which must be checked before the survey is started.

Statutory undertakings or bodies will be able to furnish information about their services in and around the site. Although this information may not be entirely reliable, at least it should be a good guide to what to look for during the surveying operations.

The local authority will most probably have the original plans submitted for its approval at the time of building. However, later additions or alterations not requiring permission will not be shown, neither will amendments which took place during construction; nor will siteworks inaccuracies be noted. At best, copies of these drawings will save sketching time; yet they will nevertheless give a feel of the property before survey work begins and they should be acquired if at all possible.

7.8 General

The whole process of building survey is almost the building process in reverse. Whereas an architect produced working drawings from which the building was erected, the surveyor has the building and he produces drawings which, in their way, are a reproduction of the original working drawings.

It is from such drawings that 'other work' is created and upon which 'other skills' are based. Just as a building relies upon its foundations for its stability, this 'other work' relies upon accurate and full survey drawing(s) for its success.

There is no short cut to competent building surveying. The principles have to be learned, and followed, if the work produced is to be worthwhile.

7.9 Code of Measuring Practice

7.9.1 The purpose

The purpose of this code is *to provide succinct and accurate measuring definitions* for use when

    i) describing land,
    ii) specifying building works,
    iii) undertaking conveyancing,
    iv) planning a property sale,
    v) formulating lettings particulars, etc.

The code is published jointly by *The Royal Institution of Chartered Surveyors (RICS)* and *The Incorporated Society of Valuers and Auctioneers (ISVA)* and aimed at the adoption of a common understandable surveying 'language'. All dimensions, areas and volumes are calculated in a uniform manner.

7.9.2 Scope

The code deals only with *measurement practice*. Associated valuation aspects, such as the zoning of areas for shopping, or certain types of buildings where particular methods of definition are required, e.g. licensed premises needing areas defining as 'non-public', 'ancillary', etc., are not covered by the code. However, certain definitions within the code may also have general meanings, for example, *Site area* and *Gross site area*.

7.9.3 Definitions

The code embodies the following *definitions* and typical *recommended applications*.

a) General Definitions
    i) *Gross External Area (GEA)* which was formerly termed 'gross floor space', deals with the measurement of a building taking each floor into account *to include*

- perimeter wall thickness including any projections.
- area of internal walls and partitions as occupied in plan.
- columns, piers, chimney-breasts, stairwells, etc.
- lift rooms, plant rooms, tank rooms, etc.
- open-sided covered areas ⎱ to be identified
- enclosed car-parking areas ⎰ separately.

        but *to exclude*

open balconies,	open fire escapes,
open 'covered' ways,	open vehicle parking,
domestic outside W.C.s, etc.	

The recommended applications for the use of GEA include
    the rating of houses,
    site coverage – plot ratio,
    estimation of building costs.

    ii) *Gross Internal Area (GIA)* which is generally as (i) above, but *excluding* external wall thicknesses.
        The recommended applications include the estimation of building costs.
   iii) *Nett Internal Area (NIA)* which was formerly referred to as 'effective floor area', is the *useable space measured to internal finish* of structural, external or party walls of each floor but *excluding*

- toilets and lobbies, cupboards, etc.,
- all areas excluded in (i) GEA,
- essential access areas, e.g. halls, landings, etc., including corridors (e.g. fire and circulation),
- space occupied by permanent apparatus such as heaters, ventilation plant, etc.,
- car parking areas,
- floor space with a headroom of less than 1.5m (see Fig. 7.12),
- for shops – recessed entrances and arcade displays.

The recommended applications for use of NIA include
    property valuations,
    industrial rating,
    flats, but not other types of domestic property.

    iv) *Gross External Cube (GEC)* used for *lateral measurement* (as GEA) *multiplied by the height of the area*. Fig. 7.13 shows methods of determining vertical measurement for

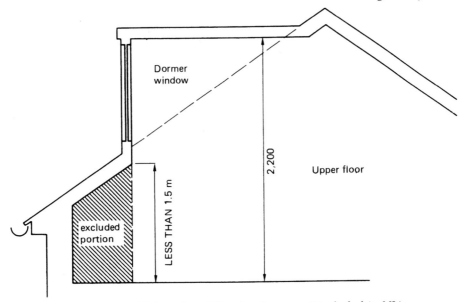

Fig. 7.12 Floor spaces with less than 1.5 m headroom not included in NIA

- strip and raft foundation
- flat and pitched roofs.

The recommended applications include estimation of building costs, planning purposes, etc.

v) *Plot Ratio* the ratio of the GEA to the *Site Area* where the site area is expressed as one, e.g. 3.5:1.

Note: (a) *Site Area* is the total area of the site within the site title boundaries, measured in a horizontal plane.

(b) *Gross Site Area* is the Site Area *plus* any area of adjoining roads, enclosed by extending the side boundaries of the site up to the centre of the road, or to 6m out from the frontage, whichever is the less. *GSA* would be used when designing residential developments and calculating planning densities.

The code defines other methods of site measurement which are graphically explained in Fig. 7.14. In addition, the code deals specifically with *shops* and *residential measurement*. Briefly these are

b) Shops (see Fig. 7.15)

i) *Sales area*, the NIA useable for retailing purposes;
ii) *Storage area*, the NIA used for storage not part of (i);
iii) *Shop frontage*, overall external frontage including entrance but excluding recesses at doorways;
iv) *Overall frontage*, gross measurement across the front of the building;
v) *Shop width*, internal surface to surface across width;
vi) *Shop depth*, as above but using depth.

The recommended applications include estate agency and valuation practice.

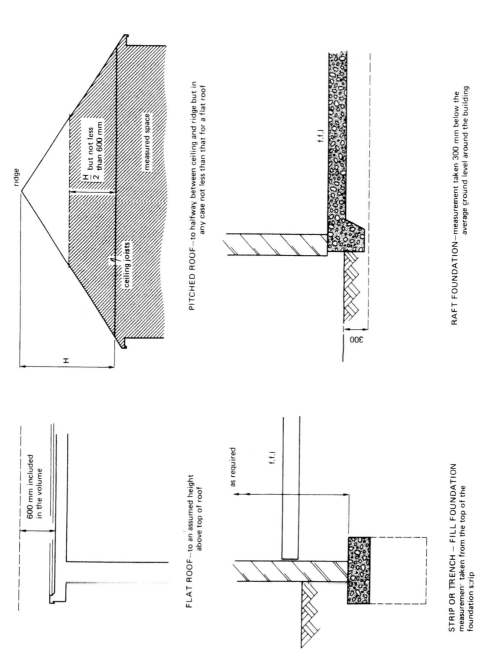

Fig. 7.13 Methods of determining vertical measurements for GEC calculation

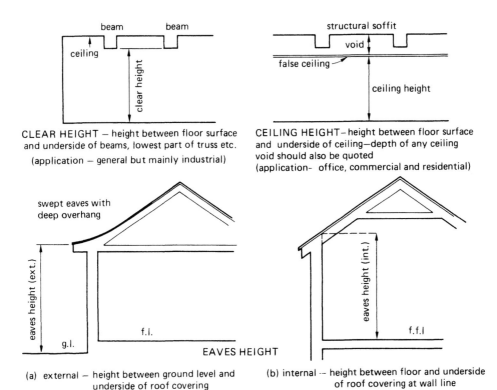

CLEAR HEIGHT – height between floor surface and underside of beams, lowest part of truss etc.

(application – general but mainly industrial)

CEILING HEIGHT– height between floor surface and underside of ceiling–depth of any ceiling void should also be quoted
(application– office, commercial and residential)

EAVES HEIGHT

(a) external – height between ground level and underside of roof covering

(b) internal – height between floor and underside of roof covering at wall line

(application – general in both cases)

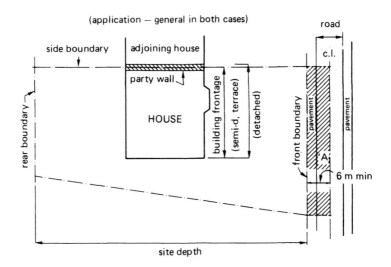

SITE AREA – total area within side, rear and front boundaries (application – general)

GROSS SITE AREA – as above plus area 'A' i.e. adjoining roads in line with side boundaries, up to centre line of road or 6 m from frontage (whichever is the greater) (application – C.P.O. and residential)

Fig. 7.14 Definitions of site measurements

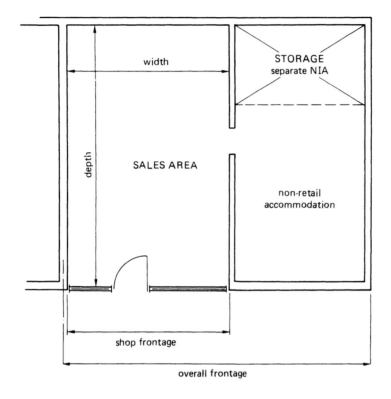

(a) SHOPPING

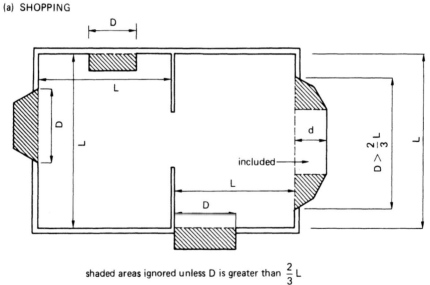

shaded areas ignored unless D is greater than $\frac{2}{3}$ L

(b) RESIDENTIAL

Fig. 7.15 Measuring for shops and residences

c) **Residential** (see Fig. 7.15)
  i) *Generally* all measurements are taken to internal wall finishes or back of cupboards. It is normal to ignore chimney breasts and bay windows, either as an addition to, or deduction from, the area being calculated unless they exceed two thirds the length of the wall in which they are placed. If bay windows are to be included, always work to the greatest depth.
  ii) *Hallways* are excluded unless, because of their unusually large size, they are used as habitable areas.
  iii) *Kitchens* are measured to internal surfaces *above* worktops. If cabinets are permanently fitted and occupying useable floor space this must be stated.
  iv) *Bathrooms* are excluded unless they are unusually large.
  v) *Garages* are measured overall internally. Projecting piers etc., should be ignored unless they comprise severe restrictions to the internal area.
    The recommended applications include estate agency and valuation practice.

7.9.4 The Appendix

to the code deals with principal UK Statutory Definitions used in Measuring Practice specifically in relation to

Town and Country Planning Acts, and
Finance Acts

Schedules and paragraphs are quoted from various Acts of Parliament which recommend methods of measurement for certain classes or types of buildings. Obviously, this text cannot cover the code in too great a detail and the student is recommended to refer to the actual code for specific information.

Exercises on Chapter 7
1. List and fully describe *four* reasons for which existing buildings are measured.
2. What are meant by the terms 'separates' and 'separates and overall' when used in conjunction with the survey of existing buildings?
3. How would you check the thickness of (a) an internal wall and (b) an external wall? Use sketches to illustrate the principles.
4. Explain how you would measure the section through a two-storey property.
5. Discuss in detail how you would proceed to survey a Victorian detached house in its own grounds. List the sequence of operations, equipment, etc. that you would use.
6. List *five* purposes for which the 'Code of Measuring Practice' would be used.
7. Define the terms 'Gross Internal Area' and 'Nett Internal Area'. Explain the difference.

8

Setting-out

In Chapter 1, *setting-out* was defined as 'the term used for the operations necessary for the correct positioning of proposed works on the ground and their dimensional control during the construction process'.

Following the completion of the survey operations, the information will be used by the designers of the proposed works to determine the best way that the site can be developed in order to satisfy the wishes of the client or developer. Once design agreement has been reached, working drawings, bills of quantities, and other tender documents are prepared and eventually *a contract to build* is signed.

The pre-build process has gone full circle, because the builder will now require copies of the survey information, particularly the *grid of levels* and the *setting-out drawing(s)*. These will form the basis of the building operations, and without them the builder cannot start on site.

In traditional building, no great accuracy is required because of the type of materials and building techniques which are employed, and the odd 5 or 10 mm either way will not create very many problems. However, advances in technology, in terms of new materials and methods of construction, have determined a fresh set of rules whereby precise setting-out is often necessary.

Precisely dimensioned frames and factory-made components require great care and attention being paid to the setting-out operations. Loose fit can be very expensive to put right, and the contractual battles which will surely ensue can be very lengthy, adding even more expense to what probably started out as a simple straightforward contract.

8.1 Setting-out drawings

These are prepared by the architect or engineer and from them the contractor physically *sets out* the development on the ground. Main wall lines, corners of buildings, centre lines of stanchions, centre lines of sewers, etc. are related to *fixed points of reference* by *dimensional co-ordination*. All dimensions are written on the drawing and shown in such a way that the buildings or artefacts can be positioned by normal survey techniques. The positions of any T.B.M.'s should also be shown, along with their values.

The fixed points of reference and the T.B.M.'s are important in that they may be used throughout the building operations for purposes of checking and controlling the works. A typical setting-out drawing is shown in Fig. 8.1.

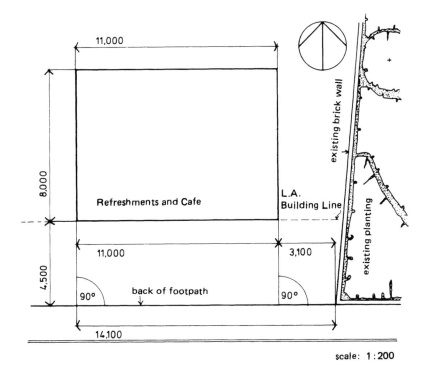

11,000

8,000

Refreshments and Cafe

L.A.
Building Line

existing brick wall

existing planting

11,000

3,100

4,500

90°

back of footpath

90°

14,100

scale: 1:200

Fig. 8.1 Setting-out drawing

8.2 The setting-out operation

This is the everyday task of the construction team. On small works it may be done by the foreman and checked by the architect, although the author has on occasion, as the architect, set out individual houses with the help of his own assistant and the builder. On larger works, the contractor's *site engineer* or *resident engineer* would be responsible for setting out the works, but this does not absolve the architect from his responsibility for checking that the setting-out is correct, although the contractor is responsible for the accuracy of the work. The client's *clerk of works* will no doubt also be involved in the procedure.

It is also important that there is liaison between the person who carried out the original land survey, the person who prepared the setting-out drawing(s), and the person who will actually be in control of the operations on site. The one who did the original survey will know the site as well as, if not better than, anyone else, and to include him in the early stages of the project may well pre-empt problems arising out of site conditions.

The student must appreciate that the setting-out operation is an important phase in the contractor's programme and will have an effect on his site organisation. It may not be possible to set out the site as one operation – indeed, it may not be desirable where work is being carried out in phases. Yet, just as the smooth running of the contract will depend on delivery dates and the like, so will the early stages depend on the setting-out being accomplished without delay. The operation must be able to be carried out accurately and in its proper sequence, with due consideration having been given to demolitions, clearance, and the provision of the proper information necessary to the work before the contract starts.

8.3 Requirements and conditions

Setting-out is a three-part problem, which is to say that whatever is being set out must be

i) in the correct position,
ii) at the correct level, and
iii) erected vertically.

For all three parts of the problem to be resolved satisfactorily, certain requirements and conditions must be met by the personnel carrying out the operations and the drawn information from which they will work.

8.3.1 Personnel

Personnel who are engaged in setting-out operations must have

a) a good knowledge of the principles and techniques of surveying;
b) a modicum of common sense, in order to deal with any unusual problems which may arise; and
c) a sound knowledge of construction technology (i.e. building construction).

Operatives must be able to translate information given on drawings into fact on the ground and assess whether the given information is sufficient. If it is not, they must be capable of explaining where the deficiency lies and what further information is required so that the operation can be continued or even started.

8.3.2 Drawings

The person preparing the setting-out drawing(s) must carefully consider the amount and type of information to be given. A drawing with *too little* information will cause the operations to be stopped until further advice is given, while a drawing with *too much* information can have the same effect by causing confusion on site. In other words, the drawn information must be just right if the operation is to run smoothly. Figure 8.2 illustrates the point.

8.4 Equipment

The normal surveying equipment as described in earlier chapters will be used, along with the following additional items described in this section. The nature and complexity of the building or engineering work will determine the accuracy that needs to be achieved, which in turn will determine which pieces

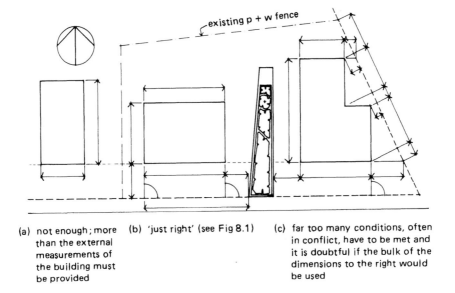

(a) not enough; more than the external measurements of the building must be provided

(b) 'just right' (see Fig 8.1)

(c) far too many conditions, often in conflict, have to be met and it is doubtful if the bulk of the dimensions to the right would be used

Fig. 8.2 The amount of information given must be 'just right'

of equipment will be selected for the task, e.g. dumpy level of theodolite, fibreglass tape or steep tape, plumb-bob and line or optical plummet, and so on.

8.4.1 The Cowley level

This instrument, shown in Fig. 8.3, differs from other forms of level in that there are neither lenses nor a bubble-tube in its construction. To quote the Clarkson Group catalogue for the Watts SL200 Cowley Level: 'Far more precise than the use of string, wire, straight-edge, or liquid level and much faster than a dumpy level, the "Cowley" is the perfect low-cost instrument for all levelling purposes. It combines simplicity with speed and accuracy. The "Cowley" is always ready for use, involves no setting-up, and requires no adjustment.'

It must be pointed out to the student that in spite of the above recommendation, there is no likelihood of the "Cowley" supplanting the ordinary forms of level, although it can have many uses in the levelling of building works. It is ideal for working in tight spaces and for checking differences in floor height; it can be used to level brickwork courses by substituting a stand for the tripod; and, by fitting a special attachment, the level can be used for setting out gradients.

The instrument is used in conjunction with a metal or timber staff, 1.5, 2, or 3 m in length, which has an adjustable cross-bar. The staff is graduated on its back and, unlike an ordinary levelling staff, is *read by the staffman* and *not* by the observer.

The instrument has a pendulum mirror system which defines a horizontal line of collimation when the instrument is approximately levelled on its tripod. This mirror system is in two parts, as can be seen from Fig. 8.3(b), with

(a)

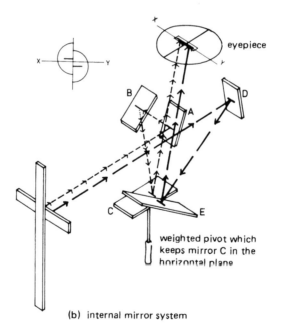

eyepiece

weighted pivot which
keeps mirror C in the
horizontal plane

(b) internal mirror system

Fig. 8.3 The Cowley level

the left-hand side composed of the two mirrors noted as D and E. The right-hand side has three mirrors – A, B, and C – and it is mirror C which is mounted on a pivoted pendulum so that it will remain horizontal irrespective of how the instrument is set. Because the mirror systems are set side by side, two images of the target, separated by a vertical line, are presented to the eyepiece which is to be found on top of the instrument.

The observer signals the staffman to move the cross-bar up or down until the two images in the eyepiece are symmetrical and meet at the central dividing line. When this condition is reached, the cross-bar is at the same height as the optical centre of the instrument and the *staffman* reads the distance from the ground to the bar shown by the graduated scale on the back of the staff.

The accuracy of the instrument is claimed to be ±5 mm over a sight of 30 m, and this is improved over shorter ranges. The range of sight is limited to about 30 m because the eyepiece has no system of magnification, although by using a *distance target* the range may be increased.

8.4.2 The site square

This is an optical device for setting out right angles whereby unskilled labour can attain an accuracy of ±5 mm in 30 m or 1 in 4800 (approx).

The instrument is basically *two telescopes* mounted one above the other and with their lines of sight set at 90° to each other (see Fig. 8.4). The site square is supported on a tripod which can be set up over a fixed mark on the ground. The *lower telescope* is then aimed along the line from which the right angle is to be established, being brought to bear on any site mark in the line by moving the telescope (a) in the vertical plane and (b) laterally, by means of a fine-tuning screw. Once the adjustment of the lower telescope is complete, the *upper telescope* will trace out a line at right angles to the original line and a further site mark can be positioned as required by moving this telescope in the *vertical plane only*.

8.4.3 Ancillary equipment

a) **Pegs** are normally made from either of two materials:

i) *Timber* – usually 50 mm × 50 mm section, of variable length but having a pointed end to facilitate driving into the ground. A timber peg may have a *nail* fixed into its top at the centre, to locate exactly the station point.

All *setting-out pegs* should be clearly marked with a 50 mm deep band of red paint round the top, while levelling pegs should have a band of blue paint. In both cases, additional information stating to what the peg refers – e.g. centre line, face of wall, manhole number, R.L. = x etc. – should be marked on the side of the peg in paint or waterproof marker crayon etc.

ii) *Steel* – usually formed from lengths of steel reinforcement rods, cut to suitable lengths and carefully driven into the ground. Once their position has been checked, it is normal to carefully surround them with

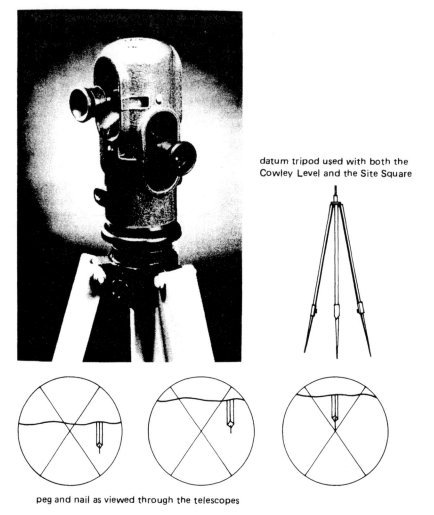

datum tripod used with both the
Cowley Level and the Site Square

peg and nail as viewed through the telescopes

Fig. 8.4 The SH 50 site square

concrete. Identification marks may be scribed into the surface of the concrete before it sets.

All site personnel should be instructed in the use of the pegs – which are which etc. – before the works have proceeded very far, because misunderstandings at this stage can be serious and costly in the later stages of the job.

b) Lines may be of hemp, string, nylon, or wire. The weather conditions will play as big a part in the selection of material as the nature of the job. Lines must not stretch and sag through rain, or become so taut that the pegs or *profile boards* become displaced or the line itself breaks.

c) Profile boards are used in conjunction with pegs so that extended line positions may be marked. By using profile boards, the string or wire lines can be removed in the knowledge that when they are required again they can be positioned exactly as they were originally.

Figure 8.5 shows the use of pegs, lines, and profile boards for setting out a simple right-angled corner.

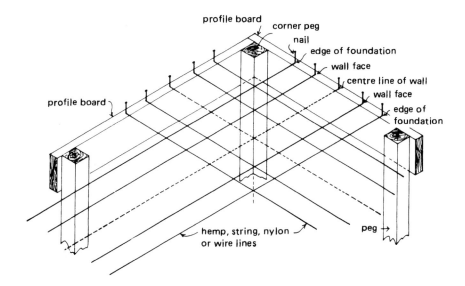

Fig. 8.5 Use of pegs for setting out a right-angled corner

8.5 Plan control

This is taken to mean the control of the *plan* or *horizontal dimensions* and the *shape* of the structure under construction. Imaginary lines representing *centre lines*, *face lines*, *base lines* etc. are established over the site and on the proposed alignment of column centre lines, wall face lines, and so on. Sooner or later these lines are bound to be obstructed by the construction process in the form of plant, spoil heaps, stacks of bricks and blocks, and all the other bits and pieces which go to make up a building site. It is customary, therefore, to establish a *reference frame* of accurately located lines *outside the construction* and *clear of all obstructions*. This means that setting-out pegs, profile boards, and lines should be outside the excavation area, which will allow constant checking to be carried out as the work progresses.

8.5.1 Setting out a simple traditional building by tapes

A simple traditional building, such as a house, will be located by means of either *two base lines* or *one base line and a fixed point*, with the setting-out achieved solely by the use of tapes. The base line may well be the building line as defined by the local authority, or alternatively the road line or the back of a footpath if these are straight lines.

A *setting-out drawing* will be provided by the architect or engineer which will show the building in relation to the selected base line and contain the measurements to be set out, in written form. If the student now refers back to Fig. 8.1, it should not be too difficult to understand the statement made in Chapter 1 that 'setting-out is the converse of surveying', in that the setting-out drawing contains the 'picture' to be put on to the ground by means of survey techniques.

From the setting-out plan and other working drawings, the contractor will determine where to put his pegs so that they will be outside the excavation. The building in Figs 8.1 and 8.6 has simple cavity walls of 275 mm width on 575 mm wide strip foundations, so if the pegs are placed 500 mm from the external wall faces they will be safe from disturbance by the building operations, from the excavation of the foundation trenches onwards.

Procedure   The sequence in which the pegs are placed is as follows, and should be read in conjunction with Fig. 8.6.

i) Fix pegs A and B on the base line.
ii) *Calculate* the diagonal from peg B to peg C and set out the distances BC (18.060 m) and AC (13.500 m) using *two steel tapes* in order to locate peg C.

Although most 'local' builders would attempt to set out the right angle at peg A by means of the '3–4–5' rule, it is more accurate and faster if the calculation of the third side of the triangle is carried out.
iii) Position peg D by the same method as above, but setting out the right angle from peg B.
iv) Pegs E and F are then lined in or measured, as appropriate.
v) Calculate the diagonals CE and DF and check that these measure as calculated (15.000 m).

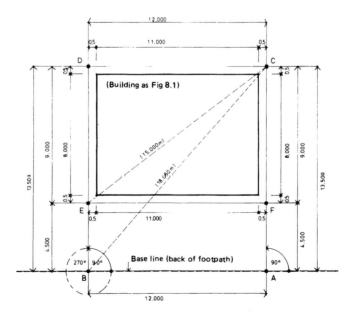

Fig. 8.6 Sequence of placing pegs

Pegs C, D, E, and F are *the main setting-out pegs* and it is around these that the profile boards will be erected (see Fig. 8.5). Pegs A and B are merely the subsidiary pegs necessary to the initial setting-out and, once the work has been checked, these pegs may be removed.

vi) By using a level and staff, a mark can be placed on the main pegs C, D, E, and F at finished floor level or perhaps at some multiple of 150 mm above or below the floor level which will tie in with the coursing of the brickwork.

vii) The profile boards can then be erected level with this mark and thereby act as *sight rails*.

The positions of wall face lines, trench lines, etc. are marked on top of the profile boards by gently tapping nails into the top surface of the boards, their positions having been measured from the corner peg.

viii) Hemp, or other, lines are tied between opposite pairs of profile boards, and it is a simple matter for the site operatives to transfer these lines vertically down to the ground by means of a builder's spirit-level or a plumb-bob and line.

Where the building is not a simple rectangle, it is nevertheless treated as such initially in terms of the overall dimensions, with profile boards being erected accordingly. Subsidiary profile boards are then erected for any 'returns' and 'set-backs' as shown in Fig. 8.7.

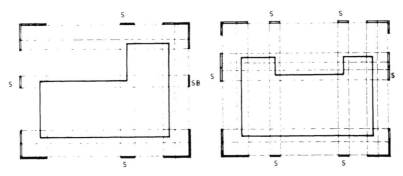

S = subsidiary boards for 'returns' and 'set-backs'

Fig. 8.7 Use and positioning of profile boards

8.5.2 Use of linear and angular measurement

While the method of setting out the simple traditional building relies on the use of written dimensions and the ability to set out right angles and measurements solely by the use of tapes, more complex work, which requires a greater accuracy, makes use of *optical angle-measuring equipment*.

Primarily, their use will be for setting out the right angle, and the instruments used will be either the site square for small works or the theodolite for more involved structures. Linear measurement will still normally be done by steel tape, although special circumstances such as large bridge construction, deep foundations, etc. may necessitate the use of electronic distance-measuring (E.D.M.) equipment.

The principle can be easily explained by referring to Fig. 8.6, for which the setting-out sequence when using a theodolite would be as follows.

    i) Set up the instrument over peg A and sight along the base line. Measure out 12.000 m along the line and fix peg B.

    ii) Turn the instrument through 90°, measure the distance 13.500 m on the line, and position peg C.

    iii) Set up the instrument over peg B and orientate on to peg A. The instrument is then *turned through 270°* and 13.500 m is measured along the line to fix the position of peg D.

    iv) Check by measurement that the distance from peg D to peg C is 12.000 m.

    v) Set up the instrument over peg D and orientate on to peg C. Sight on to peg B and check that the angle CDB is 90°.

    vi) Pegs E and F will be fixed on their respective lines by measurement.

The lines containing pegs A, F, C and B, E, D could be set out at right angles to the base line by means of the site square, the peg positions being located by measurement. It would then be necessary to calculate the diagonals AD and BC and EC and DF and check by measurement that they measure 18.060 m and 15.000 m respectively.

8.5.3 Use of a two-line reference frame

For larger and more complex structures it may not be practicable to simply use tapes for setting out the work. Similarly, working to profile boards set outside the excavation may be impossible. In such cases, optical instruments such as the site square and the theodolite will be used.

In Fig. 8.13(a), two *reference lines* have been set up which are at right angles to each other, one being the base line as defined by the local authority building line. Because only two lines are being used, lines of string or wire cannot be used for locating the wall lines etc. and it is necessary to set up a theodolite at the intermediate pegs a, b, c, and d and set out the required elements by *direct sighting and measurement*.

8.5.4 Setting-out a steel-framed building

Although the steel-framed building looks a most complex problem, the setting-out is relatively straightforward in spite of the greater accuracy that is required than for simple traditional building.

The contractor has the responsibility of setting out and pouring the foundations and bases upon which the steelwork will be fixed by the specialist subcontractor, and it is quite normal for tolerances of ±3 mm or better to be required for such contractor's work. It is necessary, therefore, that the setting-out is accurately executed by theodolite and steel tape.

A *four-line reference frame* would most certainly be used, as shown in Fig. 8.13(b). The four stations, A, B, C, and D, are carefully positioned on the ground by means of the theodolite and steel tapes, being marked by semi-permanent pegs. Other points, a, a_1, b, b_1, c, c_1, etc., which mark the ends of the centre lines of rows or lines of stanchions, are lined in along the reference frame by the theodolite and their positions are carefully measured and pegged. When the survey work has been completed and all the necessary pegs have been placed in position, it is a simple task for the site operatives to locate the centre point of any stanchion by means of lines placed across appropriate opposite pairs of pegs.

Similar methods may be used to fix the face lines of walls, while a line may also be fixed at a specified distance away from a wall or row of columns so that elements may be related to the line by means of offsets.

Although reference lines are normally at right angles to each other, the principles of extending lines and measuring will still apply where other angles are necessary due to the shape of the structure.

8.5.5 Curve ranging

Because cars and trains have a wheelbase (i.e. the distance between the centres of the front and rear wheelhubs) they cannot pivot or turn through 90°. This is why footpaths at road/street junctions have 'rounded corners' and why road and railway tracks curve in order to change direction. It is also necessary to curve roads because long sections of straight highway are not only potentially dangerous (due to monotony leading to the tiredness of car drivers) but also because they are considered, by some authorities, to be aesthetically poor.

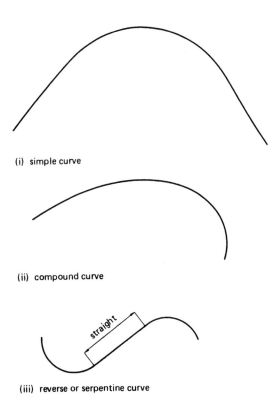

(i) simple curve

(ii) compound curve

(iii) reverse or serpentine curve

Fig. 8.8 Types of curve used for roads and tracks

The four major curves used in the design and layout of roads and tracks are (see Fig. 8.8):

a) **Simple curve** – usually part of the circumference of a circle the radius of which depends upon the given conditions.

b) **Compound curve** – formed of a succession of arcs of different radii.

c) **Reverse or Serpentine curve** – composed of two or more arcs, adjoining, but having their centres at opposite sides. In a 'reverse curve' a straight line should always intervene between the curved portions.

d) **Transition curve** – specifically inserted immediately following a straight section to introduce, gradually, the centrifugal force experienced when travelling along a circular path at speed. It begins with an infinite radius which gradually reduces until the radius of the required simple curve is reached. It returns to the new straight by a second transition curve.

An easy to understand example of such a curve is on a railway track where, in order to allow the train to take the bend at speed, the outer rail is elevated so as to avoid too sudden a rise in centrifugal force. Roads and racing circuits are 'banked' round curves for the same reason and this 'banking' is referred to as *superelevation*.

The principle fundamental to the setting out of all curves is that the straight lines into which they flow must be tangents to the curve. This applies to both *horizontal curves* and *vertical curves* – the basic difference between these being that

i) horizontal curves are commonly formed by arcs of circles, termed *circular arcs* or *circular curves* and are curved lines which lie throughout their lengths in the horizontal plane, while,

ii) vertical curves between straight tangents are formed using parabolic curves, since it is rare for the two tangents on either side of the intersection to slope at the same angle.

It is fairly obvious that in building works, smaller curves may be set out by means of a *full-size template* e.g. an arch, or from points on the ground located after the calculation of co-ordinates, e.g. a garden wall. However, for setting out roads, footpaths, kerbs, railway tracks, centre-lines, etc., these simple methods are not always practical and other methods have to be used.

A curve is set out in the form of either

i) a series of very short chords which, for practical purposes, together form a curve, or

ii) a series of straight lines, successive lines being joined by large diameter circular arcs of which the 'straights' are tangents.

a) **Geometry of the circular curve** which must be understood before a study can be made of setting out methods. Fig. 8.9 shows two straight lines AB and DE (representing the centre-lines of two straight roads) which when produced (i.e. extended) intersect at point C. The 'straights' are connected by an arc such that the lines ABC and EDC are tangential to the circle of which the arc (BdD) is a part, the tangents meeting the circle (arc) at points B and D. Fig. 8.9 also defines the geometry together with the terms normally used for the parts in practice.

The use of elementary trigonometry enables the lengths of the chords, tangents and angles to be determined easily. For example, consider the triangle OBC in which the angle OBC is a right angle.

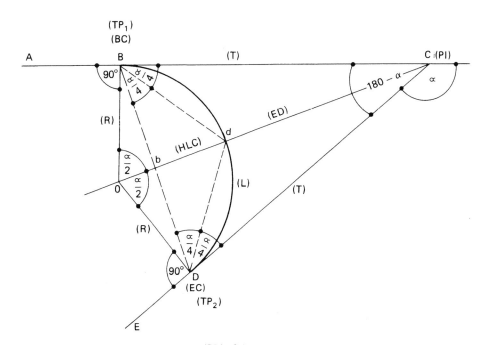

Cd is *the external distance* (ED) of the curve
BD is *the long chord* (LC)
bd is *the height of the long chord* (HLC)
B is a *tangent point (TP₁)* and the *beginning of curve (BC)*
D is a *tangent point (TP₂)* and the *end of curve (EC)*
BC and DC *are tangents (T)* to the curve
C is *the point of intersection* (PI) of AB and ED produced
        being also termed *the apex of the curve*
BCD is *the apex angle* = 180 − x *(also termed intersection*
        angle)
BOD is *the deflection angle* = 180 − apex angle = x
CBD is *the total tangential angle* = CDB = ½ deflection angle
BO is *the normal to the line AC* and is also *the radius* (R)
        of the curve as is DO, the normal to the line EC

Fig. 8.9 Geometry of the circular curve

(1) $\text{Tan} \dfrac{\alpha}{2} = \dfrac{T}{R}$

   $\therefore T = R \tan \dfrac{\alpha}{2}$; $R = T \cot \dfrac{\alpha}{2}$

(2) $\dfrac{Cd + d0}{R} = \sec \dfrac{\alpha}{2}$

   but since Cd = ED and d0 = R

$$\frac{ED + R}{R} = \sec \frac{\alpha}{2}$$

$$\therefore \quad ED + R = R \sec \frac{\alpha}{2}$$

$$\therefore \quad ED = R \sec \frac{\alpha}{2} - R$$

$$ED = R \left(\sec \frac{\alpha}{2} - 1\right)$$

Note: The quantity $(\sec \alpha - 1)$ is a function termed the *exsecant* of angle α. Since $\sec \alpha$ equals

$$\frac{1}{\cos \alpha}$$

the use of either exsecant or cosine tables is appropriate.

(3) $\qquad \cos \dfrac{\alpha}{2} = \dfrac{Ob}{R}$

but $\qquad Ob = R - db$

$$\therefore \quad \cos \frac{\alpha}{2} = \frac{R - db}{R}$$

$$R\cos \frac{\alpha}{2} = R - db$$

$$\therefore \quad db = R - R\cos \frac{\alpha}{2}$$

i.e. Height of Chord $= R\left(1 - \cos \dfrac{\alpha}{2}\right)$

Note: The function $(1 - \cos \alpha)$ is termed the *versed sine* of angle α (see Chapter 2, Section 2.6.2(d)). If *versine* tables are not to hand, use cosine tables appropriately.

(4) The arc (BdD of length L) subtends a central angle α at 0.

$$\frac{BdD}{\text{perimeter of circle}} = \frac{\alpha}{360°}$$

$$\therefore \qquad \frac{L}{2\pi R} = \frac{\alpha}{360}$$

$$\therefore \qquad L = \frac{2\pi R\alpha}{360°}$$

Note: α must be stated in *degrees and decimals of a degree* **not** in degrees, minutes and seconds. In order to save calculations, the use of tables of *circular measure of angles* for R = 1, is recommended.

(5) Because triangles BCO and CDO are congruent triangles and have one common side, it follows that

$$\text{angle BOD} = 180° - \text{angle BCD}$$

Hence (i) *Deflection angle = 180° − Apex angle*

$$\text{angle CBD} = 90° - \text{angle DBO} = \text{angle BOC} - \tfrac{1}{2}\text{ angle BOD}$$

Hence (ii) *Tangential angle = half deflection angle.*

Similarly, if Bd be any other chord, angle CBd is the tangential angle for that chord and equals half the deflection angle for that chord, i.e. $\tfrac{1}{2}$ BOC.

Note: Deflection may be *right* or *left* and is determined by observing the finishing point from position O and looking towards point C. If the finishing point is to the right of the line of sight the deflection is right and if on the left, the deflection is left.

(6)   In triangle BbO

$$\sin \frac{\alpha}{2} = \frac{Bb}{R}$$

$$\therefore Bb = R \sin \frac{\alpha}{2} \ ; \ R = \frac{Bb}{\sin \dfrac{\alpha}{2}}$$

Since Bb is half the long chord, it follows that the

$$\textit{long chord length} = 2R \sin \frac{\alpha}{2}$$

Note: Where *chord tables* are used, remember that

  i) the length will be listed opposite the central angle

$$\alpha \ (\textit{not} \ \frac{\alpha}{2} \ \text{as above) for R = 1}$$

  ii) the table entry must be multiplied by R.

(7) Again in triangle BbO

$$\cos \frac{\alpha}{2} = \frac{Ob}{R}$$

$$\therefore Ob = R \cos \frac{\alpha}{2} \ ; \ R = \frac{Ob}{\cos \dfrac{\alpha}{2}}$$

b) Setting out small radius curves up to 30m (see Fig. 8.10) can be accomplished using a steel tape. The procedure is simply

  i) Locate the centre of the circle of which the curve forms a part and mark with a peg (and nail);
  ii) Hook the tape on to the nail and extend to the required radius;
  iii) Either trace out on the ground the required curve or place pegs equidistant around the circumference of the circle at distances apart of about 15°, i.e. $\tfrac{1}{4}$ radius (see Fig. 8.10a).

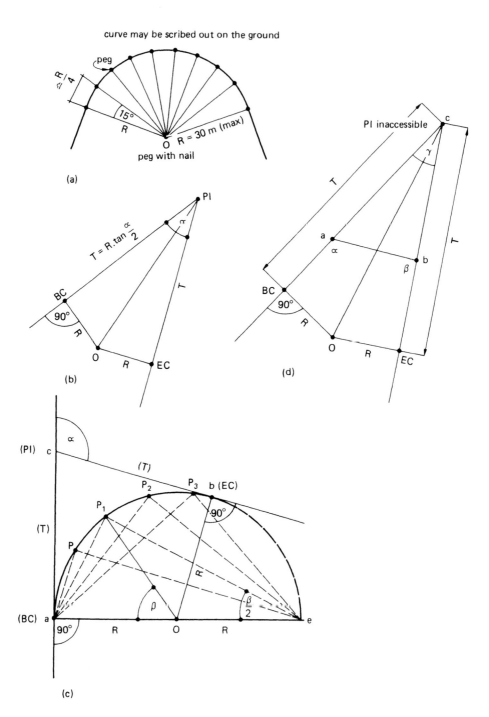

Fig. 8.10 Setting out small radius curves

Assuming that *the point of intersection (PI)* also termed *Apex of the curve* is accessible, having marked out the 'straights' on the ground and being given the radius R, an alternative procedure is:

i) Locate the PI and set up the theodolite over this point.
ii) Measure the intersection angle α shown in Fig. 8.10(b).
iii) Calculate the tangent distance (T) using the formula $T = R \tan \alpha/2$ (Equation 1).
iv) Set out a distance T along the straights from the PI to establish the 'beginning of curve' (BC) and the 'end of curve' (EC) marking these points with pegs.
v) Set out a perpendicular to the tangent at BC (or EC) and measure a distance R along it, marking with peg and nail to fix the centre of the circle.
vi) Hook the tape on to the nail in the peg and swing the required arc.

The use of the Optical Square (see Fig. 8.10(c)) is an appropriate method of setting out a curve where the diameter is free of obstruction. The procedure is:

i) Locate the 'straights' and the tangent points *a* and *b* and mark in the normal way.
ii) Locate and fix the centre 0.
iii) Set out the diameter *a0e* with points *a* and *e* each marked with a ranging rod.
iv) Locate points P, P_1, P_2 .... on curve.

Any point on the curve can be located by setting out a right angle such as aPe, aP_1e, etc., since the diameter of a circle subtends an angle of 90° at the circumference of the circle. A rod is placed approximately on the curve and an optical square, through which both *a* and *e* are sighted, held against it. The rod may be moved about until the images of the ranging rods at *a* and *e* coincide, at which point, the hand-held rod defines a point on the circle.

Normally the length of sight of the optical square is limited to 30m (see Section 2.6.1(a)) and for fairly small curves or rough setting out, the hand-held prism or even the cross-staff may be used.

It is also possible to set out a specified chord distance such as aP_1; this can be calculated using the formula

$$aP_1 = 2R \sin \frac{\beta}{2}$$

and if it is required to divide the arc into n even chords, the angle $B = \dfrac{\alpha}{n}$

However, in practice, *the apex of the curve may well be inaccessible* and it is impossible to set up the theodolite over the PI in order to measure the intersection angle. Another procedure must then be followed. This is:

i) Select two points on the straights, *a* and *b*, visible from one another, and so that the line *ab* can be measured (see Fig. 8.10(d)).
ii) Set up the theodolite at each of these points in turn and measure the angles α and β as shown in Fig. 8.10(d).

If the student now considers the triangle *cab*.

$$angle\ cab = 180° - \alpha$$
$$angle\ cba = 180° - \beta$$

also *angle acb* $= 180° - (180 - \alpha + 180 - \beta)$

∴ *apex angle* $= \alpha + \beta - 180$

The triangle *cab* can now be solved since all three angles and the length of side *ab* are known.

Thus: $\dfrac{ca}{sin(180-\beta)} = \dfrac{ab}{sin\ \gamma}$

∴ $ca = ab \times \dfrac{sin(180-\beta)}{sin\ \gamma}$

and $\dfrac{cb}{sin(180-\alpha)} = \dfrac{ab}{sin\ \gamma}$

∴ $cb = ab \times \dfrac{sin(180-\alpha)}{sin\ \gamma}$

iii) Knowing the radius, the tangent distance T can be calculated from

$$T = R\ tan\ \dfrac{\alpha}{2} \quad \text{(Equation 1)}$$

where $\alpha = (180° - \gamma)$ = deflection angle.
Since $aBC = T - ac$
and $bEC = T - cb$

it is a simple matter of calculation to enable tangent points BC and EC to be located and pegged out.

iv) Set out the curve by any appropriate method.

Another problem is *when the sight-line from the tangent points BC or EC to the centre 0 is obstructed.*
In this event the procedure is:

i) Bisect the angle between the straights (180 − deflection angle).
ii) Calculate the distance ED (Fig. 8.9) using the formula

$$ED = R\ (sec\ \dfrac{\alpha}{2} - 1) \quad \text{(Equation 2)}$$

iii) Measure a distance of (R+ED) from the PI along the centre-line PIO to locate the centre of the circle.
iv) Set out curve as appropriate.

c) **Setting out curves of radius greater than 30m** requires a different approach and several methods are available. The methods are also appropriate where the centre of the circle is inaccessible due to obstructions of one sort or another.

Method 1. Deflection distances (see Fig. 8.11(a)) is a method which relies on linear measurement only with little calculation necessary. The method is also useful for a 'trial run' and the procedure is:

i) Locate the straights and peg out one or both tangent points.
ii) Assuming a chain length ℓ and a curve radius of R, calculate the values of

$$\frac{\ell^2}{2R} \text{ and } \frac{\ell^2}{R}$$

iii) From point a(BC) and with one end of the chain fixed on it, pull the chain taut and swing the free end until it is at a point b, such that the perpendicular distance from the tangent bb_1 is equal to

$$\frac{\ell^2}{2R}$$

Mark with peg.

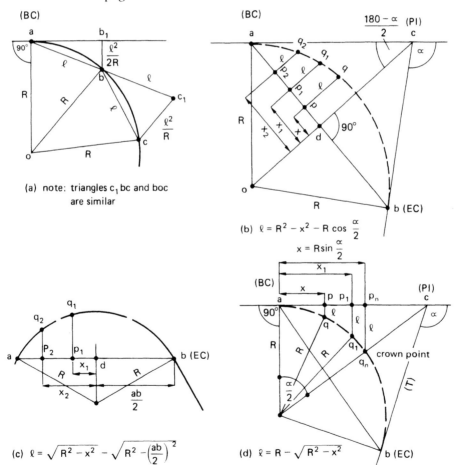

(a) note: triangles c_1 bc and boc are similar

(b) $\ell = R^2 - x^2 - R \cos \dfrac{\alpha}{2}$

$x = R \sin \dfrac{\alpha}{2}$

(c) $\ell = \sqrt{R^2 - x^2} - \sqrt{R^2 - \left(\dfrac{ab}{2}\right)^2}$

(d) $\ell = R - \sqrt{R^2 - x^2}$

Fig. 8.11 Setting out curves of radius greater than 30 m

iv) From point b, with one end of the chain fixed on it, pull the chain taut so that abc_1 is a *straight line*. Mark point c_1 with a peg.

v) Using a second chain with one handle fixed on point c_1 swing the free end of the chain, keeping it taut, until it is at point c, with the distance cc_1 exactly

$$\frac{\ell^2}{R}$$

by calculation. Mark point c with a peg or pin.

vi) Continue round the circumference locating further points as for point c (stages (iv) and (v))

Note that in order to set out the exact distances bb_1, cc_1, etc., a second chain or steel tape should be used. The student should have no difficulty in proving the method, using similar triangles and that

$$c_1c = \frac{bc^2}{ob} = \frac{\ell^2}{R}$$

is an exact value for the standard deflection distance.

Method 2. Offsets from the long chord (see Fig. 8.11(b)) is another method which not only uses linear measurement but also requires the deflection angle α to be measured and a slightly more complex calculation to be carried out. The procedure is:

i) Locate the straights ac and bc marking the tangent points a and b.

ii) Sight the long chord ab and fix its centre point d marking it with a peg with nail. This is best done by bisecting the *intersecting angle acb* although d could be set out by direct measurement.

iii) Set out points q, q_1, q_2,......, on the curve by means of perpendicular offsets, from the long chord at points p, p_1, p_2, ...

Any point on the curve can be located in this way by fixing distances x, x_1, x_2 ...... along the chord from its centre and calculating the offset length $\ell = R^2 - x^2 - R \cos \dfrac{\alpha}{2}$, again easily proved.

If the centre of the curve is inaccessible and the deflection angle cannot be measured a curve can be set out by offsets provided the positions of the tangent points a and b and the radius R are known (see Fig. 8.11(c)). The procedure is:

i) Locate and mark tangent points a and b.

ii) Set up a string line between these two points and mark the centre of the chord d.

iii) For distances x_1, x_2, x_3 .... along the chord from its centre, calculate the offset length

$$\ell = \sqrt{R^2 - x^2} - \sqrt{R^2 - \left(\frac{ab}{2}\right)^2}$$

Method 3. Offsets from the tangent (see Fig. 8.11(d)) is a linear method requiring chain/tape and *optical square* and, prior to electronic calculators and latterly computer programs, tedious calculations. However, it is an acceptable method and the procedure is:

i) Locate the straights and tangent points *a* and *b* and mark in the normal way;

ii) Extend the straights towards the point of intersection;

iii) Set-off distances x, x_1, x_2 ......, along one tangent to give points p, p_1, p_2 ......

iv) Set-off a perpendicular offset from each point, p, p_1, etc., of length $\ell = R - \sqrt{R^2 - x^2}$ to give points q, q_1, q_2 ...... on the curve.

Note: The *crown point of the curve* is reached when $x = R \sin \dfrac{\alpha}{2}$

at which stage the procedure is repeated commencing at the other tangent point.

d) Setting out curves of very large radius may be done using any of the methods already explained. However, for long curves of very large radius as used for major roads and railways, it is more appropriate to use a theodolite to set out the curve by means of deflection angles.

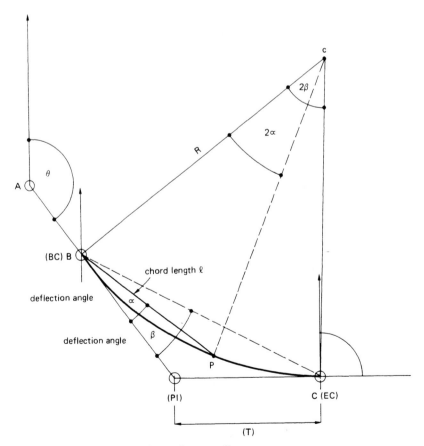

Fig. 8.12 Setting out curves of very large radius

Although this type of work is outside the scope of this text, the principle is fairly easily understood. Fig. 8.12 shows points A,B,C, of which the co-ordinates would be given. The radius R would also be known. In principle the procedure is as follows:

    i) From the co-ordinates of A and B, calculate the whole circle bearing (WCB) θ.

    ii) Choose a convenient chord length ℓ which will depend on the type of work being undertaken, e.g. for retaining walls $\ell = 2-5m$, for kerbs $\ell = 5-10m$.

    iii) Calculate *deflection* angle $\alpha = \sin^{-1}\left(\dfrac{\ell}{2R}\right)$

    iv) Calculate WCB of point P from B $= \theta \pm \alpha$;
       It can be easily seen that in this example it would be a minus.

    v) Stages (ii), (iii) and (iv) are repeated for other points on the curve. However, for tangent point c, the chord length ℓ is calculated from the co-ordinates of B and C.

    vi) Erect the theodolite over point B and set out points on the curve.

For a long curve or where sights will be obstructed, it will be necessary to set out the curve in sections which will require the theodolite to be moved for each section. Further points to note are that if the PI is given rather than the co-ordinates of C it is necessary to

a) Calculate the tangent distance T from the co-ordinates of B and PI;

b) Calculate the deflection angle $\beta = \tan^{-1}\dfrac{T}{R}$

c) Calculate chord length $\ell = 2R\sin\beta$.

8.5.6 Other aspects of plan control

a) Permanent points Where a building is to be demolished to create a site for a new structure, the cleared site will present a far different picture to that presented before demolition took place. Very few, if any, of the original features will remain visible and identifiable. It will be of great assistance to the setting-out of the new structure if a number of fixed *permanent points* are located during the original survey and are carefully preserved. By this means, continuity will be maintained and the setting-out drawings can be based on known control points.

b) Use of original traverse In a city centre, where a building is to be demolished and replaced, the original survey may well have been based on a *traverse* set up in the surrounding streets. The original traverse-survey lines can be used as reference lines to which stanchion centres and wall lines etc. can be related (Fig. 8.13(c)).

c) Use of a grid On large sites being developed for housing or industrial estates, the use of a grid will enable the individual buildings to be located more easily and accurately than if a single base line is used for the entire site.

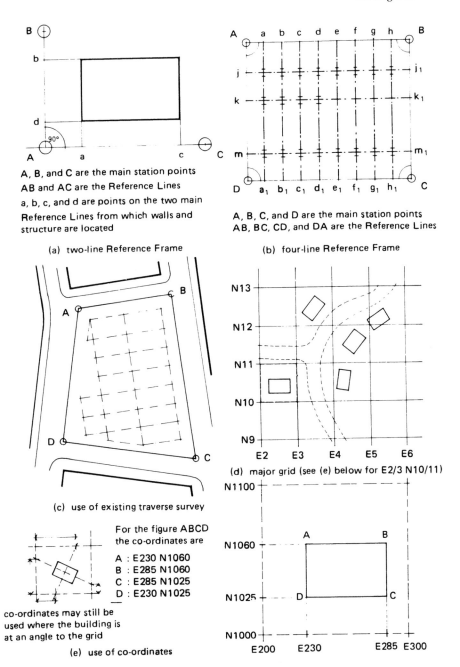

A, B, and C are the main station points
AB and AC are the Reference Lines
a, b, c, and d are points on the two main
Reference Lines from which walls and
structure are located

(a) two-line Reference Frame

A, B, C, and D are the main station points
AB, BC, CD, and DA are the Reference Lines

(b) four-line Reference Frame

(c) use of existing traverse survey

(d) major grid (see (e) below for E2/3 N10/11)

For the figure ABCD
the co-ordinates are

A : E230 N1060
B : E285 N1060
C : E285 N1025
D : E230 N1025

co-ordinates may still be
used where the building is
at an angle to the grid

(e) use of co-ordinates

Fig. 8.13 Use of reference frames, traverse, grids, and co-ordinates to aid setting-out

The use of letters and numbers to identify the grid lines will enable setting-out drawings to be produced for isolated areas of the site, provided that they can be related to a master sheet and none of the grid lines pass through any of the buildings.

d) Use of co-ordinates   In some cases it may be appropriate to apply the techniques of co-ordinate survey to the work of setting-out. Again the technique is used in reverse, in that co-ordinates are given in order to locate the point as opposed to knowing where the point is and fixing it by finding its co-ordinates. The reference-frame lines (which may in fact be grid lines as discussed in (c)) can be given arbitrary north–south, east–west meridians which will enable them to serve as *co-ordinate axes*. Figure 8.13(e) shows a building within a 100 m grid with the corner positions fixed by co-ordinates. The grid would be set out on site and, using line N1000 as a base line, the points would be set out as described in Section 8.5.2. When setting out the grid on site, permanent points would be established at the grid stations.

If the building is at an angle to the grid pattern, co-ordinates may still be used to locate the positions of the corners of the building but *angle bearings* may also be given. It is possible to set out points by angle bearings and distances related to one control point, or by angle bearings given from two or more control points, as in very precise work.

e) Plan control on upper floors   Where a building consists of more than one floor, control of the upper storeys can be maintained by plumbing *upwards* points which have been fixed at ground-floor level. This is very often done in conjunction with *vertical-alignment control* (see Section 8.7).

8.6 Height control

Provided that accurate semi-permanent T.B.M.'s have been set up on the site, the control of heights or levels should be relatively simple. T.B.M.'s should always be levelled from Ordnance Survey B.M.'s and should be set up at points where they are unlikely to be abused, even to the extent of being protected on busy sites.

Ordinary levelling techniques are used to establish and check the levels required in the construction, and the instruments already described will be found on all building sites at some time or other, particularly during the early stages of the contract.

Where levels of points above the line of collimation need to be set up or checked, this may be done by using the inverse-levelling technique, as described in Chapter 5. It is interesting to note that where an inverted level needs to be taken that is higher than the length of the staff, a white steel tape may be suspended from the point and held taut so that it can be read through the instrument in the usual way.

8.6.1 Setting out a T.B.M.

Applying the principle of working from the whole to a part, *a levelling-control system* can be put into effect by first establishing a *master benchmark* (M.B.M.) from which the site may be covered by a number of T.B.M.'s. On small sites the M.B.M. and the T.B.M. may be one and the same, with no other T.B.M.'s required.

The M.B.M. will be established by means of *flying levels* from an O.S.B.M., with the levelling being closed within agreed tolerances.

8.6.2 Setting out a level from a T.B.M.

The process is very similar to locating a contour, using the height-of-collimation method of calculating and booking. The following procedure should be read in conjunction with Fig. 8.14(a).

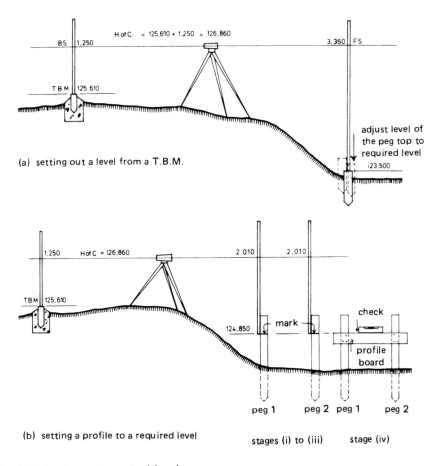

(a) setting out a level from a T.B.M.

(b) setting a profile to a required level

stages (i) to (iii)   stage (iv)

Fig. 8.14 Setting out required levels

i) Let the T.B.M. have a known R.L. = 125.610 m and the level required be 123.500 m.

ii) The instrument is set up midway between the T.B.M. and the required point, to equalise the B.S. and F.S. distances.

iii) A B.S. reading is taken on the T.B.M. as 1.250 m. The R.L. of the line of collimation is

$$125.610 \text{ m} + 1.250 \text{ m} = 126.860 \text{ m}$$

For a level of 123.500 m, the required staff reading will be

$$126.860 \text{ m} - 123.500 \text{ m} = 3.360 \text{ m}$$

iv) A peg is driven into the ground at the required point and the staff is placed on it and read. The level of the top of the peg is adjusted until such time as the required staff reading is given when the foot of the staff is placed on top of the peg. The top of the peg will then be at the required level of 123.500 m.

The same process is adopted when transferring a level or a series of levels from one part of the site to another, although the process must be checked by levelling back onto another point of known height.

8.6.3 Setting a profile to a required level

The same procedure is adopted as for the previous operation and is illustrated in Fig. 8.14(b).

i) Let the required level be 124.850 m and the height of collimation, as before, be 126.860 m. The required reading on the staff will be

126.860 m − 124.850 m = 2.010 m

ii) The staff is placed adjacent to peg 1 and adjusted until the required reading can be seen at the centre cross-hair of the instrument. The peg is then marked by a line level with the foot of the staff.

iii) The same procedure is followed for peg 2.

iv) The profile board is then lightly tapped into position with its top aligned to the marks on the pegs and checked for horizontality by the builder's spirit-level before being fully secured to the pegs.

v) The top of the board may be checked by placing the staff on it and checking for the required reading.

8.7 Vertical-alignment control

This aspect of control can be divided into two distinct parts:

i) the transfer, vertically, of control points or lines to higher or lower levels;

ii) the provision of vertical control lines and the checking of the verticality of building elements.

8.7.1 Vertical transfer of control points and lines

Once reference frames or a series of control points have been established at ground-floor level, one of two things may be done:

i) the original reference frames may be constructed at the new level, or

ii) fixed points may be transferred to the new levels so that new reference frames can be constructed from them.

The accuracy required depends on the work under construction, but basically three methods are used for the vertical transfer of positions (see Fig. 8.15): plumb-bob and line, optical plummet, and theodolite.

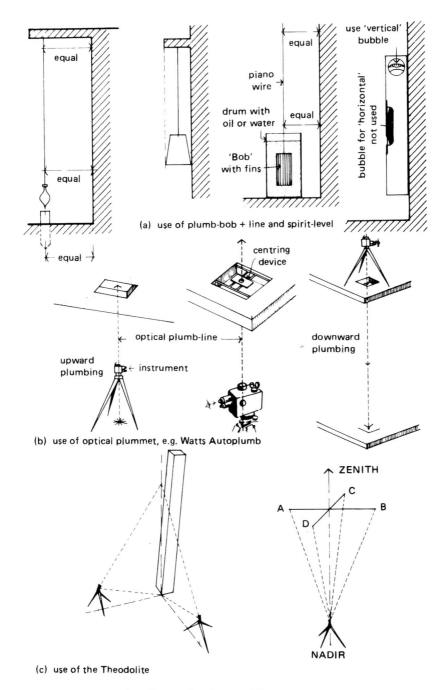

(a) use of plumb-bob + line and spirit-level

(b) use of optical plummet, e.g. Watts Autoplumb

(c) use of the Theodolite

Fig. 8.15 Vertical transfer of control points and lines

a) **Plumb-bob and line** (Fig. 8.15(a)). This is a traditional method suitable only for small buildings of one or, at most, two storeys. The line is very prone to movement by the slightest wind and requires shielding even on quite short lengths. Lines may be string, nylon, or piano-wire – depending on the length required to be plumbed. The bob is slow to settle and there are various ways of overcoming this, such as using a bob with fins and suspending it in a drum of oil or water. Care must be taken to ensure that the bob is not resting on the bottom, and to this end it is probably better to use a glass container.

Over very short distances, rather than setting up a plumb-bob and line, it is simpler and quicker and equally effective to use the builder's level which is fitted with two bubbles, one of which is for vertical plumbing.

b) **Optical plumbing**   The transfer of points by optical instruments may be up or down. Mine shafts can be plumbed as easily as a ground point can be transferred up through a tall building.

Optical plummets can be separate instruments or can be built into other instruments such as theodolites. The plummet can be used to position a theodolite *exactly* over a ground mark and then transfer that mark to a higher plane, such as the underside of floor slabs, and various devices can be used in conjunction with the plummet when transferring a point vertically from one level to another. While older instruments relied on a bubble system to achieve verticality, many modern instruments use the automatic-prism system found in automatic self-levelling equipment.

There are many different types of instrument on the market – so many that to discuss the subject further would go outside the scope of the present text, although the student should note their existence with future studies in mind.

c) **Vertical transfer and alignment by theodolite**   It is possible to relocate a ground point at a higher level by using a pair of theodolites. A vertical line is the intersection of *two* vertical planes. Such a plane is described by a theodolite when it is transited. It follows that, if a ground point is observed by a theodolite from two different directions in plan (not necessarily at right angles to each other), the intersection of the collimation lines for the two directions will relocate the point. Two theodolites are normally used simultaneously, for the sake of speed and efficiency.

One of the restrictions to the use of the instrument in this type of work is due to the space required around the building.

Because the transit axis is not horizontal, instrumental error occurs which increases with elevation. This necessitates face-left and face-right work to minimise the error, with a maximum elevation of 22°. By fitting a diagonal eyepiece the maximum elevation can be improved.

8.7.2 Vertical control and checking for verticality

Normal building practice is to build vertically, except in special or exceptional circumstances, and some method of ensuring that walls and columns are vertical must be employed during the building operations.

a) Vertical control This makes use of the plumb-bob and line. When the line has settled and is known to be correctly positioned, the lower end may be made fast with the line taut. A *visible and physical* line has thus been provided from which the site operatives can measure offset distances as they are required. If the control is carried out by the optical plummet, no visible line is shown and an instrument man would need to be brought in to check any new work.

Other methods may be devised for specific works in particular situations, and much ingenuity has been, and will continue to be, seen on a building site.

b) Checking for verticality Obviously the plumb-bob and line can also be used for plumbing vertical faces, as can the builder's spirit-level.

Higher-accuracy checks will require the use of the theodolite or optical plummet. It has already been stated that the telescope of a theodolite defines a vertical plane when it is elevated or depressed. This fact can be used to check the verticality of a row of columns by setting up the instrument carefully along the offset line used to set out the centre-line positions of the columns. A graduated staff is then held horizontally at the foot of the column and the reading is noted. If the staff is then held horizontally against the top of the column, the same reading must be given for the column to be vertical.

8.8 Excavation control

Excavations or *earthworks* are carried out in order to change the levels of existing ground or to dig trenches for underground services. Typical operations are

a) to excavate an area to a new, lower level;
b) to fill over an area to a new, higher level;
c) to level a sloping area by means of cut and fill, whereby the higher part of the area is dug out and the spoil is deposited on the lower part so that the whole area eventually has the same R.L.;
d) to excavate for cuttings and to fill for embankments to accommodate a route for a road or railway etc.;
e) to excavate land to form slopes;
f) to excavate trenches to receive pipes and cables, following which the trench is *back-filled* to either existing or new levels.

Excavations may vary from a simple site strip of the vegetable matter, say 200 mm, to bulk excavation many metres deep, but all need to be controlled in such a way that correct levels are achieved. Before methods of controlling the actual excavations are discussed, several definitions of simple control equipment need to be given.

8.8.1 Definitions

a) A sight rail is very similar to a profile board, being simply a horizontal strip of wood, 50 mm to 100 mm deep, fixed to two upright supports which are driven into the ground on either side of the line of the proposed excavation.

The level of the bottom of the excavation is calculated and the top of the sight rail is set at a specific height above this level. The chosen height of the rail should not be so low that it becomes lost in the mud or the operative needs to be on his hands and knees to use it, nor placed so high that a ladder is needed to sight from it. A practical height is between 600 mm and 1700 mm. One or more other sight rails will be erected in a similar way at other points along the line of the excavation, in such a way that the line of sight over the rails is parallel to the proposed bottom of the excavation (see Fig. 8.16).

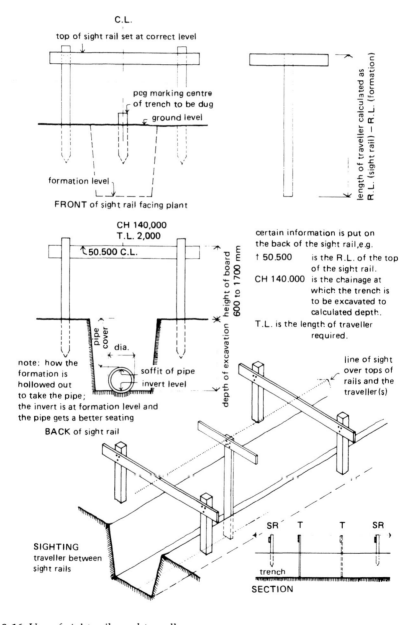

Fig. 8.16 Use of sight rails and travellers

b) A traveller, also known as a *boning rod*, is made of thin timber resembling a T-square in shape. The length overall, from the top of the cross-piece to the foot of the blade, will be the same as from the top of the sight rail to the invert, with the cross-piece being about 400 mm wide. All travellers for any one line must be the same length.

The proper depth of dig at any intermediate point between sight rails can easily be checked by sighting a traveller(s) between the rails (see Fig. 8.16). Fixed boning rods may also be placed at each end of the line, in which case they will act as sight rails. A traveller would be used between them in the normal way.

c) 'Bone' is the old-fashioned term for slope or gradient. A regular slope is still often referred to on a building site as an 'even bone'.

d) An invert is the level of the inside bottom surface of a drainage pipe. *Manhole inverts* are read on to the bottom of the channel or halfpipe which runs through the manhole. The term *invert level* is abbreviated to I.L.

e) Grade slips are used when a drainage trench is to be dug by hand. When the line of the trench has been determined, pegs are inserted at regular intervals about 500 mm from the centre line, having their R.L. marked on top. The grade slips are then marked, showing the depth to which the trench must be dug below the level of the pegs. The slips are purely for the information of the drainer.

8.8.2 Area excavation or fill

Control pegs need to be placed at regular intervals over the area. These must be carefully levelled from a B.M. or site T.B.M.

a) **To fill over** an area it is better if the pegs are placed so that their tops are at the required level; or alternatively, this may be done by profile boards and lines, sight rails, and travellers to achieve the correct finished level.
b) **To excavate** an area, the system of sight rails is to be preferred.
c) **To establish gradients**, the sight rails will be placed at the top and bottom of a slope and travellers will be used in the normal way. If the slope is steep, or if the proposed slope and the existing slope conflict, double sight rails may be necessary (see Fig. 8.17). These will maintain the practicability of sight rails. Gradients may, however, be set out directly, either by a *tilting level* fitted with a *gradienter screw* (see Section 4.5.2) or by a *Cowley level* if it has a *slope attachment* fitted (see Section 8.4.1).

Finally, it should be noted that the earthworks are carried out to the required levels *between* the lines of pegs or sight rails. These will be left on mounds of earth which will be taken away during the tidying-up process after the rails etc. have served their purpose and can be moved.

8.8.3 Cut and fill

Although the control of cut and fill is basically the same as for more general excavation work, there is a special problem attached to the formation of the side slopes. The *angle of repose* for the material being used needs to be calculated in the office. This will determine the positions of the top and bottom of the slope which require to be pegged. *Oblique sight rails* can be used to control the slope, as shown in Fig. 8.18.

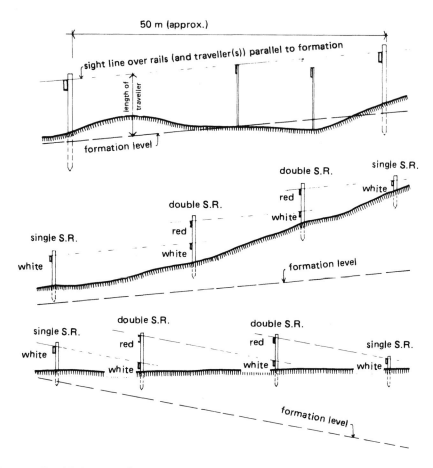

Fig. 8.17 Establishing gradients

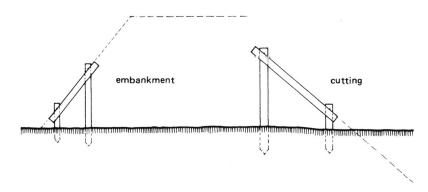

Fig. 8.18 Cut and fill

8.8.4 Trench excavation

Trench-excavation work has been virtually covered in the above text and will come under further consideration in Section 8.9. The *alignment* of the plan may be controlled by marking the overall width of the trench (equidistant about its centre line) on the sight rail. The trench centre-line mark will also be useful to the pipelayer once the trench is complete.

It may also be useful to note at this point that, when the pipelaying has begun, a traveller with a bracket of angle-iron can be used between the sight rails to check the actual inverts of the pipes as they are put down.

8.9 Setting-out and levelling for drainage work

Consideration of a small site of gentle slope will be sufficient to illustrate the principles involved.

The site will be 'gridded' and a close vertical interval will be chosen for the contouring. The student should note that, the 'flatter' the land is, the closer the V.I. must be if the drainage is to work at all without cutting deeper and deeper into the land as the *outfall end* of the trench is approached.

From the plotted contours, the floor levels of the building will be determined along with the most economical drain runs. Once the best course for the drains has been decided, the following procedure will be adopted.

i) The drain runs will be staked off on the ground from the drainage-plan measurements, by means of laths which will be placed along the centre line of the drain.

ii) Pegs will be placed at regular intervals and about 500 mm from the centre line, with grade slips if the trench is to be dug by hand.

8.9.1 Calculation of the trench depth

Before the trench depth can be calculated, the fall of the pipe must be determined. This will be done in the office by drawing a longitudinal section between the I.L.'s of the head and the outfall, showing the profile of the land over the proposed drain length. Consideration will be given to maximum and minimum permissible falls and to keeping the ground cover over the pipe to the minimum, to avoid unnecessary excavation. The fall of a pipe may vary over the full length of the run in order to save excavation, and in some cases a gentle fall to the last manhole will become a steep connection from this manhole to the main sewer.

Figure 8.19 shows a longitudinal section of a length of drain. It has been determined that the drain needs to be a minimum depth of 1.000 m below ground and the I.L. at 0 chainage (outfall) is to be 105.000 m. The pipe will be of 100 mm diameter, which suggests a rule-of-thumb maximum gradient of 1 in 40.

A gradient of 1 in 40 would result in the drain being too shallow at the peak. If the peak is given an I.L. of 106.350 m, so that it is below ground by the same amount as the I.L. at the outfall, the fall will then be 106.350 m − 105.000 m = 1.350 m in 70 m, which is a satisfactory fall of 1 in 52 (1 in 51.86 actual). (As an alternative, a fall of 1 in 60 could be used from 0 to 40 m chainage and a fall of 1 in 40 from 40 m to 70 m.)

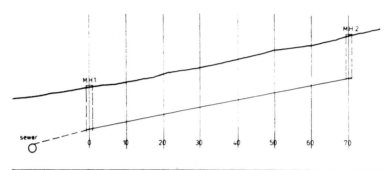

CHAINAGE (m)	0	10	20	30	40	50	60	70
GROUND RL	106,100	106,210	106,450	106,600	106,820	107,100	107,200	107,450
INVERT (pipe)	105,000	105,192	105,385	105,576	105,769	105,961	106,154	106,346*
DEPTH TRENCH	1,100	1,018	1,065	1,023	1,051	1,139	1,046	1,104
R.L. of top of 2,000 long sight rail	107,000							108,346

*note: the 4 mm difference between the assumed I.L. and the calculated I.L. of the pipe is due to the fall of 1 in 51.86 being rounded up to 1 in 52 for purposes of calculation.

Fig. 8.19 Section of drain

It is now necessary to calculate the depth of the trench at each chainage:

At 0, this will be simply the difference between the ground R.L. and the outfall I.L. of the pipe,

i.e. 106.100 m − 105.000 m = 1.100 m

At 30 m the I.L. of the drain will be higher than at 0 by (1 in 52) × 30 m,

i.e. (1 ÷ 52) × 30 m = 0.577 m

∴ I.L. = 105.000 m + 0.577 m = 105.577 m

The depth of the trench is again the difference between the R.L. of the ground and the I.L. of the pipe,

i.e. 106.600 m − 105.577 m = 1.023 m

The depth of the trench is calculated for all the chainage points and the results are shown in the table in Fig. 8.19. It is these figures which would be marked on the grade slips.

8.9.2 Calculation of the length of the traveller

Following the staking out of the centre line of the drain run, the sight rails will now be erected at each end of the run. If these are placed at a height of 900 mm above the ground at 0 chainage, the R.L. of the top of the rail will be

$$106.100 \text{ m} + 0.900 \text{ m} = 107.000 \text{ m}.$$

Since it is known that the difference between the I.L.'s over the run is 1.346 m for a gradient of 1 in 52 over 70 m, it follows that the R.L. of the top of the sight rail at the peak of the run will be 107.000 m + 1.346 m = 108.346 m. The line of sight now given by the two rails will be parallel to the required gradient of the drain.

The length of the traveller required will therefore be the difference between the R.L. of the sight rail and the I.L. of the drain at 0 chainage,

i.e. 107.000 m − 105.000 m = 2.000 m

This will be equally true at the peak,

i.e. 108.346 m − 106.346 m = 2.000 m

The student should note that by using the satisfactory fall of 1 in 52 (as opposed to the actual fall of 1 in 51.85) to calculate the rise of the pipe over 70 m, the sight rail at the peak will only be 896 mm above the actual ground level and not 900 mm as at the outfall end. The R.L.'s of the sight rails are set out as described in Section 8.6.2.

Finally, if the trench is to be dug by mechanical means, the sight rails may need to be detachable from the posts by being supported on brackets as opposed to rigid nailing. Also, the posts may have to be spread wider apart, to allow the excavator to pass between them.

8.10 Calculation of Volumes of Earthworks

In order to estimate the cost of earthworks, it is necessary to know the *volume* (i.e. quantity) of material which requires to be *dug out, graded* or *imported*. The three main methods of calculating volumes of material are

  i) by means of *cross-sectional area multiplied by length*, as used for roads, trenches, etc.,
 ii) from *contours*, and
iii) from *surface 'spot' levels* when the proposed excavation has vertical sides, e.g. a basement.

The following text deals with manual or mechanical methods of calculation which, although straightforward, can be time-consuming. They are, however, acceptable as something more than just approximations. Computer applications are discussed in Chapter 10, Section 10.2.

8.10.1 Volume from cross-sectional area

The areas of *successive cross-sections* of cut and fill are required in order to calculate earthworks volumes when preparing for the laying down of roads or similar constructions having 'length'. Cross-sections are taken at suitable intervals and the area of each obtained by one of several methods such as are described in Section 3.17.

Knowing the area of each cross-section and the distance between successive sections, the total volume may be calculated by one of three basic methods.

a) **Mean areas** where the average of all the cross-section areas is obtained and then multiplied by the overall length of the earth-works. The method is easy to apply and, although somewhat inaccurate, in that it gives *too* large a volume, it is useful for rough estimation at the initial design stage.

Let successive cross-sectional areas be A_1, A_2, A_3 ...... A_n, with the overall distance from cross-section A_1 to cross-section A_n represented by L. The volume, V, will be found from

$$V = \frac{A_1 + A_2 + A_3 + \ldots\ldots A_n}{n} \times L$$

b) **End areas and trapezoidal formula** where the total earthworks volume is treated as a *trapezoid* i.e. a solid figure with two parallel plane ends each of which is a trapezium. A typical example of such a volume is that of a cutting between a pair of cross-sections when the 'floor' of the cutting has a different gradient to the ground level.

Let two successive cross-sections be A_1 and A_2 which are a distance of ℓ apart. The volume V contained between the sections will be given by

$$V = \ell \times \frac{A_1 + A_2}{2}$$

provided that the cross-section at the midway point between A_1 and A_2 is actually the mean of them. The accuracy is still acceptable where only slight changes between successive cross-sections are in evidence.

The formula may be developed for any number of cross-sections A_1, A_2, A_3 .... A_n, becoming

$$V = \ell_1\frac{(A_1+A_2)}{2} + \ell_2\frac{(A_2+A_3)}{2} + \ell_3\frac{(A_3+A_4)}{2} \ldots\ldots \ell_{n-1}\frac{(A_{n-1} + A_n)}{2}$$

However, if the cross-sections are at regular intervals such that $\ell = \ell_1 = \ell_2 = \ell_3 \ldots\ldots \ell_n$, then the formula can be written

$$V = \frac{\ell}{2}[(A_1+A_2) + (A_2+A_3) \ldots\ldots (A_{n-1} + A_n)]$$

$$\therefore V = \ell\,[\frac{(A_1+A_n)}{2} + A_2+A_3 \ldots\ldots A_{n-1}]$$

This expression is termed the *trapezoidal formula for volumes* or *rule for volumes* and, as for 'mean areas', is not wholly accurate, giving a result that is generally *less than* the true volume.

c) **Simpson's Rule or Prismoidal formula** which is rather more accurate than methods a) and b). This method assumes that the volume of material between successive cross-sections is actually a *prismoid i.e. a solid consisting of two parallel plane end-faces (not necessarily of the same shape) with the sides joining the faces formed by continuous straight lines running from face to face.*

Let A_1 and A_2 be the areas of successive cross-sections, i.e. the end-face areas, A_M the area of the cross-section midway between the end-faces. The volume of the prismoid is obtained from

$$V = \frac{1}{6}(A_1 + 4A_M + A_2)$$

The application is more important than the proof (which will not be covered by this text) in that where numerous cross-sections have been taken, *every alternate section* may be regarded as an end-face with the 'middle cross-section' representing A_M, the distance between end-faces then being 2ℓ.

Let A_1 and A_3 be the areas of alternate cross-sections and A_2 the area of the mid-cross-section. The volume V will be found by

$$V_1 = \frac{2\ell}{6}(A_1 + 4A_2 + A_3)$$

$$= \frac{\ell}{3}(A_1 + 4A_2 + A_3)$$

Between the next set of alternate cross-sections the volume will be calculated from

$$V_2 = \frac{2\ell}{6}(A_3 + 4A_4 + A_5)$$

This can then be continued along the line of the earthworks until A_n is reached (N.B. *'n' must be an odd number*) so that

$$\Sigma V = V_1 + V_2 + V_3 + V_4 + V_5 + \ldots\ldots$$

$$= \frac{\ell}{3}(A_1 + 4A_2 + 2A_3 + 4A_4 + 2A_3 \ldots\ldots +2A_{n-2} + 4A_{n-1} + A_n)$$

This is *Simpson's Rule for volumes*, similar to his rule for areas (see Section 3.17). Points to note are that

i) each cross section appears only once;
ii) their multipliers are 1, 4, 2, 4, 2, 4, ... 2, 4, 1 respectively, and
iii) there must be *an odd number of cross-sections* just as there are an *odd number of offsets* or ordinates in *Simpson's Rule for areas*.

8.10.2 Volumes from contours

Volumes can be calculated from contour plans if the areas enclosed by successive contours are taken to be 'cross-sections' and the VI between contours as the constant distance between 'cross-sections'. The methods outlined above in Section 8.10.1 (a), (b) and (c) can be employed as appropriate to calculate the volume of material contained between any two specified closed contours. The most common application of the exercise is to calculate volumes of liquids such as water in reservoirs etc. Here each contour line on the side slope is regarded as a 'waterline' from which successive plan areas of liquid are calculated enabling volumes to be calculated. The method is also applicable to volumes of materials stocked while awaiting distribution, i.e. fill, gravel, etc.

8.10.3 Volumes from spot levels

where, as already mentioned, the excavation will have vertical sides and a level formation. A uniform grid of squares is laid down over the area to be excavated and spot levels taken at grid intersections (see Section 5.14.1). Each square on the grid can be taken as the top end-face of a *vertical prism running from formation level up to original ground level.*

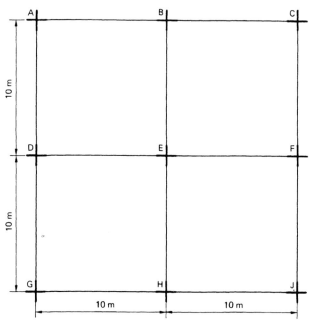

GROUND POINT	SPOT HEIGHT	FORMATION	HEIGHT (h) prism corner	No (n) SQUARES occurs in	PRODUCT (h) X (n)
A	10.00	8.00	2.00	1	2.00
B	10.25		2.25	2	4.50
C	10.50		2.50	1	2.50
D	9.50		1.50	2	3.00
E	9.75		1.75	4	7.00
F	10.00		2.00	2	4.00
G	9.50		1.50	1	1.50
H	9.50		1.50	2	3.00
J	9.75		1.75	1	1.75
				TOTAL (Σ)	29.25
				AVERAGE	7.3125

Volume = cross sectional area of one prism
multiplied by mean height of grid

From above data   V = 10 × 10 × 7.3125
= 731.25 cu.m.

Fig. 8.20 Calculation of volumes from 'spot levels'

The volume is calculated from *the cross-sectional area of each prism multiplied by the mean of its four corner heights, i.e. ground level minus formation level.* Fig. 8.20 shows the method. Although the volume of each square prism could be calculated separately, this would be a tedious and unnecessary procedure, since several of the 'corners' appear in more than one prism. For example, corner B appears in two prisms, namely, *ABED* and *BCFE*, while E appears in *four*. It is much simpler to book results and calculations as shown in Fig. 8.20 in order to arrive at *the sum of the four heights for all the squares.* This must be divided by FOUR (since each prism has four corners) in order to find *the sum of the (four) mean heights* which then only needs to be multiplied by the area of one 'prism'.

Alternatively, the grid could be interpreted as a series of triangular prisms, each having *three corners* e.g. *ABD, BDE,* etc. Each corner height would appear a different number of times e.g. B in three triangular prisms, although E will still be in *four*. E(h × n) will also require to be divided by *three* instead of four to find the sum of the mean heights prior to multiplication by the area of one triangle.

Exercises on Chapter 8
1. What are setting-out drawings?
2. List the three major requirements of setting-out.
3. Describe the Cowley level by means of sketches and notes, and explain how it is used.
4. Using diagrams and brief notes, explain how you would set out a single-cell structure in relation to a straight local-authority building line. If the walls are 275 mm cavity brickwork, calculate and show the positions of the profile boards.
5. Explain how a point may be transferred vertically, up or down, by means of (a) a plumb-bob and line and (b) an optical plummet.
6. What is a site square and how is it used?
7. List the equipment, ancillary to tapes and theodolites, necessary for setting-out operations.
8. Explain the differences between a two-line and a four-line reference frame. Where would each be used?
9. Define the following terms employed in curve ranging: long chord, tangent point, tangent distance, apex angle, intersection angle, deflection angle.
   Produce a free-hand sketch and indicate the positions of the above terms.
10. Describe the following methods of setting out curves
    a) using a chain and theodolite,
    b) using a chain and tape alone.
11. It is required to set out a curve of 64m radius, with pegs at approximately 7.5m centres. Given a deflection angle of 60°, draw up the data necessary for pegging out the curve by:
    a) offsets from the main chord,
    b) offsets from the tangent.
    Draw each operation to a scale of 1:200.
12. a) By means of diagrams, explain how a T.B.M. is set up and, once set up, how a level may be located from it.
    b) Given that the R.L. (T.B.M.) is 150.000 m and the required level is 147.750 m, make up the page of calculations necessary to setting up the level required.

13. A drain requires to be laid to a fall of 1 in 50 between manholes A and B, which are 25 m apart. If the I.L. of the lower manhole A is 101.500 m and the ground R.L. at this point is 104.000 m, and the R.L. of the ground over manhole B is 105.000 m, calculate the depth of the trench required. If the sight rail is placed 600 mm above the ground level at manhole A, calculate the length of traveller required.

9

Surveying by the plane table

The *plane table* is a most useful piece of surveying equipment which has been known to English surveyors since the sixteenth century; yet, despite the enthusiastic praise of Aaron Rathbone and others, plane-table surveying has not been used in the UK to any great extent over the years.

9.1 Definition

Plane-table surveying is a technique by which the angles or directions between survey stations or detail points are drawn directly on to a portable drawing-board in the field. Since the survey is drawn directly on to a sheet of paper as the survey proceeds, very little is required in the way of survey notes, and the office work is reduced to simply the tidying up and lettering of the field drawing.

It is a method which is in regular use in Europe, India, the USA, Canada, and other parts of the world, and it is more than likely nothing other than the unpredictability of the British climate which has made surveyors in the UK reluctant to use the plane table. In practice, however, this is not the problem it appears to be at first sight.

9.2 Uses and features of plane-table surveying

The uses are surprisingly numerous for a technique that is not favoured in the UK, and, with care, a tolerable accuracy can be obtained when the equipment is used for

a) producing maps and plans of small areas;
b) arranging a network of control points and filling in the detail between them – this can be very rapid and is most suitable for the preparation of reconnaissance and exploratory survey maps and plans;
c) filling in detail between control points which have been established by more accurate methods such as *theodolite traverse* or *theodolite triangulation* – the mapping of the detail between such control points may well be done much faster by plane table than by chain-survey technique in certain terrains;
d) the revision of existing maps and plans;
e) the location of contours;
f) the making of large-scale plans of small areas (when the equipment is used in conjunction with tacheometry techniques).

9.2.1 Methods of plane-table surveying

Basically there are *four* methods of surveying when using the plane table and its ancillary equipment:

    i) radiation,
    ii) intersection (including graphic triangulation),
    iii) traversing,
    iv) resection (also known as 'the three-point problem').

These methods may be used separately or in combination and may be extended if more complex *alidades* are used. Each method will be fully described in Section 9.6.

9.2.2 Merits of the plane-table technique

a) The plan is produced *directly in the field*, with a minimum of measurement and booking of field notes.
b) There is little chance of important features being omitted.
c) Very rapid work is possible.
d) A good field technique is fairly easily acquired by the beginner in a relatively short period of time.
e) Office work is cut to an absolute minimum, i.e. making the field plotting presentable.
f) Since complex calculations are unnecessary, the work may be carried out by relatively unskilled technicians.

9.2.3 Disadvantages of the technique

a) Work is impossible in persistently wet and/or windy climatic conditions, and it is impracticable in heavily wooded or densely bushed areas.
b) The *scale* of the map or plan *must be known before work is started*, since the actual plotting takes place on site.
c) The absence of field notes may prove a disadvantage on some jobs, especially where areas and volumes need to be calculated.

9.3 Equipment

The basic equipment is shown in Fig. 9.1 and is as follows.

9.3.1 The plane table

The plane table, from which the technique derives its name, is a special drawing-board mounted on a tripod.

For reasons of lightness, boards are usually made from softwood, although boards of mahogany or teak are also available, for which a more robust tripod

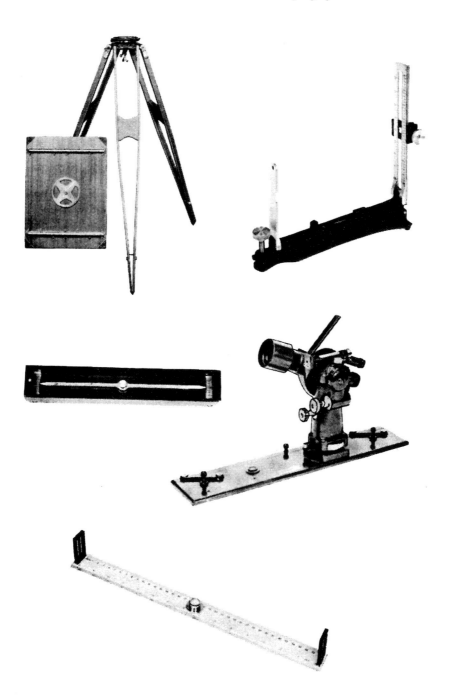

Fig. 9.1 Basic plane-table surveying equipment (*left to right from top*): board and tripod, Indian-pattern clinometer, trough compass, microptic alidade, and alidade rule

is required. The sizes of board are 360 mm × 460 mm, 610 mm × 460 mm, and 760 mm × 610 mm as standard, with a canvas carrying case. Most boards will have softwood strips on the underside to receive drawing-pins – on no account must pins be used on the surface of the board to fix the drawing-paper in position. The paper may be held in place by slide-on clips, but these can get in the way of the alidade. The advent of modern adhesives which will allow the drawing to be peeled off the board at the end of the job has enabled modern boards to have aluminium battens in place of the softwood strips. These rigid battens will render the board less likely to warp.

The tripod may be either the 'crutch' type, with fixed legs, or the 'normal' type, with adjustable legs. The author is of the opinion that the former is better because it gives greater rigidity when in use.

The board is attached to the tripod by means of a wing-nut and screw. The table may then be properly levelled in the horizontal plane by means of an adjustable head to the tripod which will have either a ball-and-socket joint or levelling screws.

When standing, with a good spread on the legs, the tripod should be about 1200 mm high, which will give comfortable working on the table. The student should note that the above items of equipment are called separately the *board* and the *tripod*, but, when set up in the field ready for work, they are together termed the *plane table*.

9.3.2 The alidade

This is a *sighting rule* made from either boxwood or metal and is approximately 380 mm in length. At either end are vertical *sighting vanes* which fold down on the rule for carrying purposes. One sight has a narrow vertical slit, while the other consists of a vertical wire stretched across an open frame. The instrument is used for pointing the direction of the distant targets and a line is ruled, parallel to each direction of sight observed, along the straight-edge of the alidade.

The Indian clinometer is a modification of the simple alidade just described, being developed during the survey of India. A pin-hole aperture is provided in the rear vane, while the forward vane, being twice as tall, has a sighting slit with marked scales of vertical angles (+ or −) and tangent values. The pin-hole aperture is in line with the zero mark on the scales, and the vertical angles (of depression or elevation) can be observed directly, as can the tangent of the angle. The instrument can be accurately levelled on the table, to facilitate the reading of the angle. Once the tangent is known, the horizontal distance to the target from the table may be measured or scaled off a map, and the difference in level between the two, or the height of a specific target, can be calculated. The Indian clinometer will enable *contouring by plane table* to be carried out.

By attaching a telescope to the basic straight-edge of the alidade, by means of a pillar-type mounting, clearer sights will be given. If a vertical circle of degrees and Beaman scales are added as well, *distances and levels may be measured tacheometrically*. Such an instrument is then termed a *microptic alidade*.

The *Beaman stadia arc* consists of two scales – the 'H' scale and the 'V' scale – mounted concentrically with the vertical circle of degrees. The 'V' scale carries values of $100 \sin \beta \cos \beta$ which may be read off and multiplied by the staff intercept s (see Section 5.3) to give the difference in elevation between the instrument and the staff. The 'H' scale carries values of $100(1 - \cos^2\beta)$ and, if this value is multiplied by s, it gives the correction to be deducted from $100\,s$ (normal tacheometry) in order to determine the true horizontal distance to the staff or target.

Example:

Staff intercept s	$= 1.750$ m
Staff reading at collimation line	$= 2.880$ m
'V' scale reading on Beaman arc	$= 24$
'H' scale reading on Beaman arc	$= 6.2$

From the above readings, the following reductions can be made:

Vertical component $= s \times$ 'V' scale reading

$$= 1.750 \text{ m} \times 24$$
$$= 42.000 \text{ m}$$

∴ change in level relative to the height of collimation $= 42.000$ m $- 2.880$ m

$$= 39.120 \text{ m}$$

True horizontal distance $= 100s - (s \times$ 'H' scale reading$)$

$$= (1.750 \text{ m} \times 100) - (1.750 \text{ m} \times 6.2)$$
$$= 164.150 \text{ m}$$

The special circle-reading eyepiece of the microptic alidade gives the angle of elevation (or depression) in degrees as well as the corresponding 'V' and 'H' Beaman-scale values, and the use of Beaman stadia arcs minimises the work normally involved in reducing the telescope stadia readings.

Although the student should be aware of the existence of the more advanced forms of the alidade, the study and use of these will again form part of more advanced studies – the first priority must be to learn how to carry out the simple form of plane-table survey. Once the principle has been thoroughly learned, understood, and practised, it is quite straightforward to extend it to the more advanced techniques.

9.3.3 A spirit-level

This is used for checking that the table is, in fact, in the horizontal plane.

9.3.4 A trough compass

This is used to orient the table in the magnetic meridian. It consists of a long needle in a shallow wooden box resembling a pencil case. The box is placed on the table and the table is rotated until the needle is central on the scales at both ends of the box.

9.3.5 A plumbing fork

This is necessary, to ensure that the station point, as marked on the paper, is *exactly* over the point on the ground that it represents.

The fork is a U-shaped piece of metal, having arms of equal length. The arm which rests on the table has a pointed end which is placed at the *drawn survey-station mark*, while the arm underneath the table has a plumb-bob and line which hangs directly under the drawn mark. When the plumb-bob is at rest over the point on the ground, the drawn mark will be exactly over this point because of the equal lengths of the arms of the fork. The fork may also be used for transferring a mark on the ground to a suitable position on the paper, e.g. at the start of a survey.

9.3.6 Drawing-paper

This needs to be of a good strong quality to resist the movement of the blade of the alidade. Certain film-type materials may also be used.

Remember that on no account must drawing-pins be pushed into the surface of the board. The paper is attached to the board in the office and, if pins are used, the paper should be turned over the edges and pinned to the *underside* of the board, using the softwood strips if they are provided.

Since the survey is plotted in the field, the usual items of eraser or French chalk are required to keep the paper clean. Pencils, steel prickers, and scale rules are also necessary items.

9.3.7 Protection

Protecting the work should be possible if the weather suddenly changes before the survey is complete.

A waterproof cover will protect the table from rain, while a large surveyor's umbrella will be useful against both rain and extreme sun. Always put the equipment in carrying cases when travelling to and from the site. When the board is detached from the tripod, it should immediately be put into its canvas case or bag, especially if the drawing is still attached.

9.4 General

Although the technique is well established, the improvements that have taken place relate entirely to the development of the alidade. As a result, the technique has been extended beyond the mere mapping or the filling in of detail for an area to *contouring* and *tacheometry*. Where these latter are combined, it is necessary to sight on to a normal staff or a special tacheometric staff placed at the distant target point, using a special alidade.

The simple form of the technique does not require the direct measurement of any angles and, with the exception of contouring and tacheometry, requires no calculations.

9.5 Fieldwork preliminaries

Just as with any other survey technique, carrying out a survey with the plane table follows a proper sequence of procedure.

9.5.1 Fixing the stations

Where the purpose of the survey is to fill in the details of a larger survey, the stations will have been fixed by means of some other more accurate technique, using an instrument such as a theodolite.

If, on the other hand, the area to be surveyed is self-contained and does not form a part of a larger survey, station points must be determined which are suited to the work in hand. This will be further discussed as the individual methods of survey are looked at in detail (Section 9.6).

9.5.2 Setting up the plane table

The proper sequence of setting up the table over a ground station point is as follows.

   i) Set up the tripod approximately over the station point so that it is firm and has a good spread on the legs.
  ii) Attach the board, with the drawing-paper already in place, to the head of the tripod.
 iii) Level up the table by manipulation of the tripod legs and/or the levelling head. Where a spirit-level is used for the final levelling of the table in the horizontal plane, it must be used in *two perpendicular directions* (see Fig. 9.2). For preliminary levelling-up, or rough working, it is probably sufficient to ensure that a round pencil will remain stationary no matter where it is placed on the table.

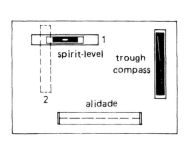

Fig. 9.2 Setting up the table: adjustable head

iv) Centre the table over the mark on the ground. This can be done by using the plumbing fork as follows.

    a) If the survey station is the first position over which the table has been set up, the mark on the ground is transferred to a suitable position on the paper *after* the levelling-up is completed.

    b) For the second (and any subsequent) station point, a mark will already be determined on the paper. This will require the table to be adjusted until this mark is *exactly* over the ground mark. The table will need to be relevelled if it has been moved.

The method of using the fork for the above operations is described in Section 9.3.5.

    Some authorities would suggest that to 'centre over the mark' is satisfied if any part of the table is over the station point. Except where the scale of the drawing is extremely small, the philosophy is not worthy of consideration. It is not too difficult to centre the table over the ground mark, and the reduction of error in the survey is well worth the effort.

v) Once the position of the station point has been marked on the table, insert a needle or a stout straight pin against which the rule of the alidade can be placed when sights are taken.

iv) Orient the table to the magnetic meridian by placing the trough compass on the table and parallel to one edge. The student should remember that *all survey drawings require a north point* drawn on them, as previously mentioned in earlier chapters.

9.6 Methods of survey

The *four basic methods* of surveying by plane table are in themselves quite simple and, once these have been understood and practised in the field, their extension to include *contouring* or *tacheometry* should not prove to be difficult.

9.6.1 Radiation

This is a method whereby a complete survey may be carried out (see Section 9.2) with one set-up of the table.

    The plane table is set up in a commanding position near to the middle of the area to be surveyed and is oriented to the magnetic meridian by means of the trough compass. The needle or straight pin is placed at point O (see Fig. 9.3), which represents the ground point of the middle of the site. The point O will obviously be approximately in the middle of the table, and sights are taken with the alidade to each important point A, B, C, D, E in turn. A pencil line, or ray, is drawn along the straight-edge of the alidade to represent the direction of each sight as it is taken.

    The distances OA, OB, OC, etc. are then measured by tape, and the measurements are plotted to scale along the appropriate ray, in order to locate the representative points a, b, c, etc.

    It is much faster and more accurate if a *telescopic alidade*, equipped with stadia lines, is used to measure the required distances from the table to the target.

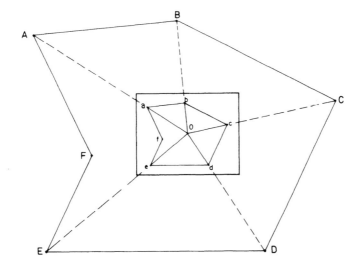

Fig. 9.3 Method of radiation

Radiation may be used to pick up local detail in close proximity to control points of a larger survey. The table may also be centred over a known station point in order to pick up surrounding detail.

By setting up at any one of the points A, B, C, D, E and orienting on to O, check sights to the important points can be taken to check the survey.

9.6.2 Intersection

This is probably the most used method of plotting detail and one by which a small area can be completely surveyed *without measurement except for the base line of the survey*.

Figure 9.4 shows the method. Two station points, A and B, have been chosen as the ends of the *base line*. These points must be intervisible and afford an unobstructed view of all the important points necessary to the survey, upon which sights will be taken from a and b, the points on the drawing which represent the stations A and B.

The distance between the stations A and B is accurately measured by tape, and this length is then plotted to scale on the paper fixed to the table, with the ends of this base line noted as a and b.

The plane table is carefully set up as described in Section 9.5.2, so that point a on the table is exactly over A on the ground, while a ranging pole is placed vertically at station B. (If the points have been pegged out, a ranging-pole stand will be necessary if the pole is to be set *exactly* over B.) With a needle or pin inserted at a, the alidade is placed carefully so that its straight-edge lies along the drawn base line ab. The table is then rotated until the sights of the alidade are trained on the pole placed vertically at station B. The table is clamped to prevent any movement of the line ab, which is now lying over the imaginary line which joins A and B on the ground.

The alidade is rotated around the needle at a, and sights are taken and rays drawn to the important points C, D, E, etc. until all the necessary points of importance have been noted.

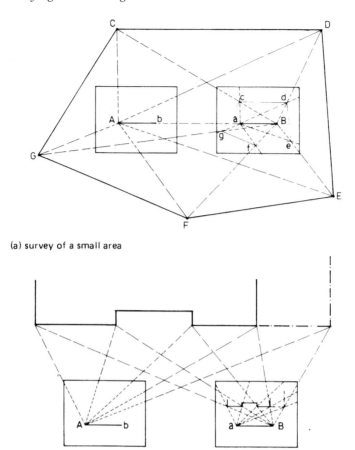

(a) survey of a small area

(b) picking-up detail

Fig. 9.4 Method of intersection

The plane table is then moved and set up so that b on the paper is exactly over B on the ground, with station A being marked with a ranging rod. With the needle at b and the straight-edge of the alidade lying along ba, the table is oriented on to station A and clamped. Sights are now taken on to the same points of importance with the rays of sights being drawn from b. Where *corresponding rays intersect*, the positions of the important points C, D, E, etc. are fixed on the drawing as c, d, e, etc., to scale and in relationship to each other.

Check rays can be drawn by setting up the table at, say, point F and orienting along the line FB. Once the survey is *proved* from this third position, it can be extended using longer base lines, such as CE or DF, in order to pick up further detail.

Where a survey is extended in order to fix subsequent station points by means of intersection, *three sightings* or rays should be used to fix the position. Triangulation in this way is known as *graphic triangulation*.

9.6.3 Traversing

This is the *graphic method* of making a traverse. It is rather tedious and is therefore seldom used unless restricted visibility prevents the use of other techniques. The method is best used for fixing survey lines and stations on the plan so that the filling-in of the detail can be done by radiation and/or intersection.

Figure 9.5 shows the procedure whereby the directions of the lines are drawn on the table instead of measuring horizontal angles or bearings. Ranging poles are erected at the traverse stations, A, B, C, D, and E and, if the starting point of the traverse is not already plotted, an arbitrary point a is marked on the table to represent ground station A. The plane table is then set up over station A, with the representative mark exactly above it. The procedure is exactly that which is adopted in the previous method – *intersection*. If the point a is already plotted, orient the table as for any base line using a to represent one end of the line; otherwise, orient by using the trough compass.

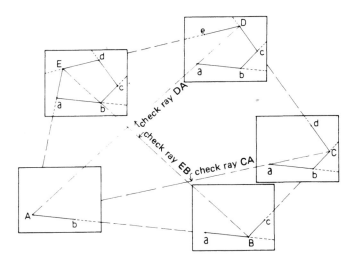

Fig. 9.5 Method of traversing

Sight the alidade on to station B and draw the ray. Measure the distance AB and plot this to scale to find point b. Set up the table at B and orient back to A. Sight on to C, draw the ray, measure BC, and plot to scale to find point c. The process is repeated at traverse stations C, D, and E.

While the table is set up at each station, the surrounding detail should be plotted by radiation or intersection, while other detail may be picked up by means of *offsets* from the lines joining the stations, as in chain-survey work. Whenever possible, *check rays* to previous stations should be plotted, e.g. CA(ca), DA(da), and so on.

If all the stations were determined before the survey began, it should not be necessary to set up over station E since this would have been sighted from both a and d. However, if its position was not determined until c was reached, or if detail around it needed to be plotted, a set-up at E would be required.

Finally, if the plotted traverse does not close, it may be adjusted by a graphic version of *Bowditch's method*. In Fig. 9.6(a), the plotted traverse eabcde$_1$ has its misclosure at station E by the amount ee$_1$. In Fig. 9.6(b), a straight line to the scale length of the actual traverse is drawn with ea, ab, bc, cd, and de representing the lines EA, AB, BC, CD, and DE. A perpendicular line ee$_1$, representing the misclosure, is drawn at one end to the same scale. The points e$_1$ and e (start) are joined by a further straight line. Perpendiculars are then erected at a, b, c, and d, their lengths aa$_1$, bb$_1$, cc$_1$, and dd$_1$ indicating the linear adjustments necessary at the plotted points a, b, c, and d of Fig. 9.6(a) in order to close the plotted traverse. These adjustments must be made on lines *parallel* to the direction of the misclosure, i.e. parallel to the line joining ee$_1$ of the plotting. The final adjusted plotting is shown by a broken line in Fig. 9.6(c).

9.6.4 Resection

This is a method normally used for fixing, or locating, an unknown position by measuring at it the angles subtended between three other points whose positions are known.

Although the angles may be measured by a theodolite or other means, the problem can be solved by the plane table. The method is usually referred to as *the three-point problem*, and three rays are drawn to three well selected points of control, already plotted on the table. This can give the surveyor more freedom of action, because he can fix his position without having to move and set up at a distant fixed point.

Resection may be done in several ways, of which the following two descriptions will suffice to explain the principle.

a) The classical solution   The art of resection lies in the selection of the three chosen points of control. The three-point problem is, in fact, insoluble when the table is on the circumference of the circle drawn through the three control points, termed *the danger circle*. When selecting points from which to attempt resection, try to have *the middle point nearest to the table*, or select points in a *nearly straight line*. For good results, the angle subtended between the two outer points should not be less than approximately 60°.

Having selected the control points, the procedure is as follows.

i) Set up the table and guess its position, marking this lightly on the drawing.

ii) Orient the table either by the trough compass or by laying the straight-edge of the alidade along the line joining the lightly marked guessed position to the most distant point of control as plotted and then sighting along this line to the actual point.

iii) Place the alidade with its edge passing through a *plotted control point* (say a in Fig. 9.7) and sight on to the actual point A. The ray is then drawn backwards, towards and past the guessed point.

iv) The above procedure is carried out for the other control points B and C.

Only by a fluke will the table be correctly oriented so that all three rays pass through a point. It is normal for a *triangle of error*, large or small, to be the result.

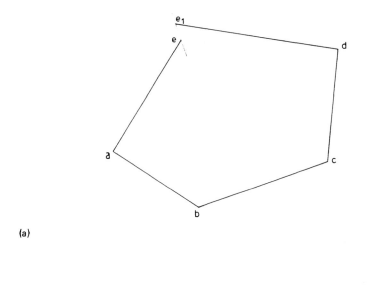

(a)

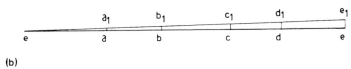

(b)

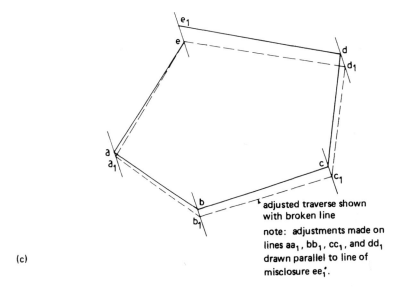

adjusted traverse shown
with broken line

note: adjustments made on
lines aa_1, bb_1, cc_1, and dd_1
drawn parallel to line of
misclosure ee_1.

(c)

Fig. 9.6 Bowditch's method of adjusting misclosure

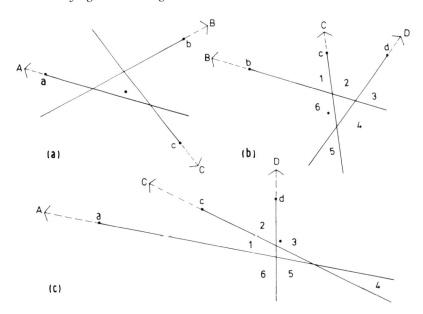

Fig. 9.7 Resection

The problem is then to *reduce this triangle to a single point*, and this is done by reorienting the table, turning it either clockwise or anticlockwise. When new rays are then drawn, it will be found that, compared with the first set of rays, the new rays will all have swung in the *same* direction, either clockwise or anticlockwise. In fact, a study of the triangle of error will show which way the rays require to be swung in order to cut at a point inside the triangle of error.

In Figs. 9.7(b) and (c), the surveyor is *outside* the triangle formed by rays from the control points and he is also outside the triangle of error. In (b), then, the rays require to be swung *clockwise* and in (c) *anticlockwise*. The rays can, in fact, be considered as being pivoted at the actual points of control.

Provided that the triangle of error is not too large, the following rules may be followed to determine the correct position from inspection of the drawing.

a) The correct position is only inside the triangle of error when the table itself has been set up inside the triangle formed by the control points (e.g. ABC in Fig. 9.7(a)).

b) When not inside the triangle of the control points, the true position will always be in sector 3 or sector 6, as numbered in Figs. 9.7(b) and (c), and will be on the same side of the ray to the most distant point as the intersection of the rays from the other two points. (For example, in Fig. 9.7(b) the true position, shown by the black dot, is adjacent to the intersection of the rays from C and D, below the ray from B, the most distant control point.)

Where control points are in a nearly straight line, sector 1 is between the left-hand and centre control points, sector 2 is between the centre and the right-hand control points, and the remaining sectors are then numbered consecutively in a clockwise direction.

c) The perpendicular distance of the true position from each ray is proportional to the length of the ray from the table to the observed point, since the rotation of the table will have the same effect on each ray.

After the first triangle of error, select a new position following the above rules. Reorient the table with the alidade laid from this new position through the most distant control point plotted and sighted on the distant station. Draw this new ray and the new rays from the other two control points. The result should be a perfect intersection or a smaller triangle of error. In the latter case, the second triangle will be the same shape as the first if the rays have not been moved far enough. If the rays have been moved too far, the triangle will be inside out. However, it should now be possible to estimate the true position of the required point, and it may also be found by joining the *corresponding apexes* of the two triangles of error.

b) A graphical solution The 'fix' may be made by using an overlay of tracing-paper as follows (see Fig. 9.8).

i) Select the ground point and drive in a peg to mark its position.
ii) Set up the table over this point (call it X) and place a sheet of tracing-paper on the board.
iii) Transfer the ground point X on to the tracing-paper, using the plumbing fork, and note this position as x.
iv) Select the three control points, which will already have been plotted, bearing in mind the points mentioned as a guide to good selection.
v) Sight on these control points, drawing rays from x.
vi) Place the tracing-paper over the survey drawing proper and adjust it until the rays pass through the corresponding points as plotted. Point x should then be pricked through on to the survey drawing and be marked up.

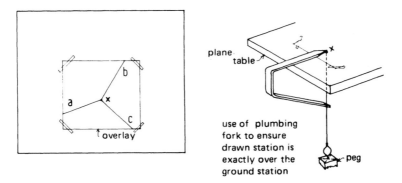

Fig. 9.8 Graphical solution

9.7 Control of plane-table surveys

Control of the survey must be effected in both the horizontal and vertical planes if errors are to be minimised.

a) **Horizontal control** is required to prevent the build-up of errors in the plan positions of stations. The required positions should be pegged on the ground and be located by means of a theodolite traverse or similar methods. Where an entire small area is to be surveyed from only two stations and one base line, errors should be minimal, provided care is taken and the procedures, as described, are followed meticulously.

b) **Vertical control** is required to minimise height errors. For work of a large scale, the plane-table stations should again be pegged and levelled by ordinary levelling techniques. This is probably better done before the plane tabling but may be done afterwards provided the pegs are not moved.

9.8 Practical applications of plane-table surveying

A small site may be surveyed as previously described from one central station point, or from two or three station points carefully tied together by accurate linear measurement. The more difficult the area in terms of detail such as trees, buildings, and other features like hedges, streams, etc., the more stations will be required.

In plane tabling, heights can be taken in two ways.

i) Set up a level adjacent to the plane table and, as each sight is made, note the reading on a staff held at the point. Using the height-of-collimation method of reducing levels, the level can be put on the drawing within a few seconds.

ii) The level may be set up over the station points after the plane tabling has been completed and, using the technique of tacheometry and a horizontal circle of degrees, the whole area can be levelled and checked in a very short time.

Direct contouring is particularly suited to the plane table if an auto-reduction alidade is used. The telescope may be set horizontally and used like a level to locate the contour, with the distance being obtained from the diagram curves. The principal of direct contouring, described in section 5.13, would be followed when working with the plane table. The advantage of contouring by plane table is that the surveyor can see immediately if his results reflect the ground on which he is working.

9.9 Accuracy and errors

With care, the results of plane-table surveying can be quite accurate, certainly for large-scale work, although the accuracy of the work will generally be governed by the scale of the plan being produced. The major error will be due to shrinkage or warping of the drawing-paper. Other errors will be due to the table not being level when sights are taken or the station, as marked on the plan, not being exactly over the station marked on the ground.

9.10 Summary

For the results to be worthwhile the student should not only practise the technique but also should follow the procedures diligently, resisting any temptation to take short cuts. Once the theory is understood, a great deal of satisfaction is to be gained from practice in the field.

Exercises on Chapter 9
1. Describe the plane-table method of surveying. Sketch and describe the equipment used.
2. In plane-table surveying, what are meant by (a) intersection, (b) resection, and (c) radiation?
3. Explain, by means of diagrams and notes, the 'three-point problem'. What is meant by the 'danger circle'?
4. Discuss the advantages and disadvantages of plane-table surveying in relation to practical applications of the technique.
5. Itemise and explain the fieldwork procedure necessary to carry out a plane-table survey.

10

Modern developments

The aim of the earlier chapters has been to provide a 'primer' for the student to whom surveying is a new topic and, at the same time, provide an easy-to-follow reference text for those who, having studied the topic, practise it but intermittently. However, it is necessary for all readers to have an awareness of modern surveying equipment and its areas of application, although no great depth of knowledge is required at this stage. This chapter serves to introduce some of the major developments in land surveying in recent years which will form the core of more advanced study.

10.1 Electromagnetic distance measurement (EDM)

The development of *EDM* has enabled distance measurement to be carried out in certain situations far more swiftly and with a greater accuracy than when conventional instruments and traditional techniques are used. Linear measurements have now become important elements of control surveys (traverses, etc.,) and the use of polar/radial methods (Section 1.4.3 et al) is quite common. The equipment may also be combined with data processing equipment (see Section 10.2).

Although still relatively expensive, for offices where the equipment will be subjected to a reasonable amount of use, it will quickly pay for itself and, no doubt, as technology develops, costs will come down. It is interesting to note that already an electronic theodolite is now cheaper to buy than the 'normal' optical type and can be more efficient. However, although at first sight electronic equipment (see Fig. 10.1) would seem to be the answer to each and every surveying problem, the student should still bear in mind that equipment must be selected as being the most suitable in terms of

    i) the work and its problems;
    ii) the accuracy required to be achieved;
    iii) the time available in which to do the work;

and so on. For some jobs, such as setting-out buildings, individual parts of construction and detail, traditional techniques can be better and will be perfectly acceptable, especially over short distances, even though the main control lines and control points have been set-out by *EDM* techniques.

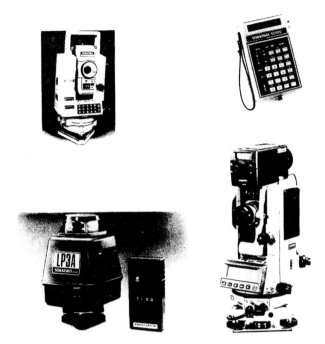

Fig. 10.1 Equipment used for Electromagnetic Distance Measurement (EDM)

10.1.1 Applications

The major advantage of EDM, apart from speed and accuracy, is that greater distances can be measured or set out irrespective of ground conditions which are often a limiting factor when using traditional equipment and techniques. Applications of EDM techniques include,

   i) measurement of critical lengths, such as long base lines, traverse lines and the establishment of control points;

  ii) accurate measurement of length where chaining and/or taping is impracticable, e.g. over water, boggy ground, chasm, wadi or donga, etc., (see Fig. 10.2);

 iii) accurate measurement of length over hilly terrain where slope calculation may not only be tedious but also introduce unacceptable errors in the final result;

 iv) setting-out or checking co-ordinates (including the co-ordinates of curves) when EDM equipment is combined with a theodolite.

10.1.2 Basic principles and equipment

EDM is based on measuring the transit time of an electromagnetic beam emitted from a transmitter/receiver to a reflecting target prism and back again.

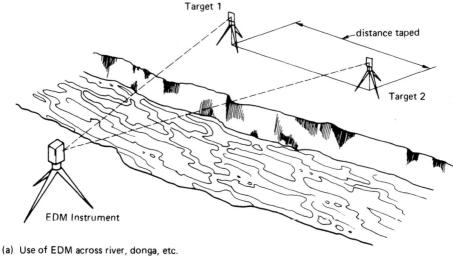

(a) Use of EDM across river, donga, etc.

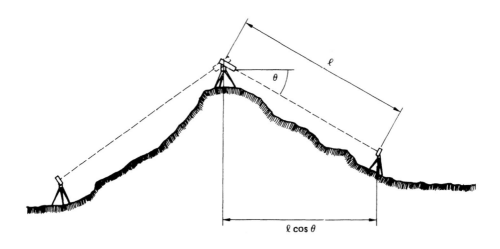

(b) Use of EDM over a hill

Fig. 10.2 EDM techniques

The direct (slope) distance is measured and displayed digitally with corrections for ambient condition applied manually or automatically in more complex equipment. Electronic 'gadgetry' will analyse the data to determine angle of slope and horizontal distance. Originally *EDM* consisted of 'add-on' units to be mounted on the telescope or the standards of a theodolite. Modern equipment is now mainly of integral manufacture and comprises

a) **Theodolite distancer** a combination of EDM and theodolite in which the *electromagnetic beam passes through the telescope optical system*. The equipment is powered by *nickel-cadmium rechargeable batteriess* and the electromagnetic signal is a beam of infra-red light. The signal is formed into a narrow beam by the lens system of the EDM but diverges slightly as distance increases. The range of the system will vary from as little as 100m in rain, mist or fog, to as much as 1000m, or more, in clear conditions, using an appropriate number of prisms to guarantee a return of signal of adequate strength.

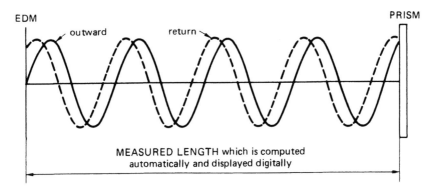

Fig. 10.3 Typical EDM signal

Fig. 10.3 shows the basic signal but a more in-depth study of how the system works by means of wavelength shift and frequency should be made by reference to more advanced texts and manufacturers' information.

b) **Target prism** is of the 'corner-cube' type, having three mutually perpendicular reflecting surfaces. It must be set towards the EDM equipment so that each reflecting face is at approximately 45° to the incoming signal which is then returned along a parallel path.

10.1.3 Fieldwork

a) Operation
The equipment is simple to operate and read and both aspects are fully covered in the manufacturers' handbooks. The procedure is as follows:

   i) Carefully set-up the EDM over a station at one end of the line to be measured. This is done in exactly the same way as for a theodolite (see Section 6.3).
   ii) Carefully set-up the prism over the station at the other end of the line to be measured. In lower order work, the prism may be set on a hand-held pole to which is fixed a 'plumbing bubble' and may be steadied using a ranging pole as a diagonal brace.

Note: In both operations, extra care must be taken to avoid centring errors.

   iii) Point, precisely, the 'aiming head' of the EDM at the prism.
   iv) Direct, approximately, the prism towards the EDM.
   v) Check signal strength and adjust as necessary until a maximum signal is returned.
   vi) Take readings and book.

Note: More than one reading of each distance should be taken. Most instruments will provide repetition of measurements, tracking, averaging and staking out. By noting each displayed measurement any 'rogue result' can be identified and dealt with. Such a reading results from the signal being interrupted or the setting of the instrument being changed during a measuring sequence.

Angles may be manually read, in which case they are observed then 'keyed-in' to the EDM whilst the distance is being measured, following which horizontal distance and vertical heights etc., will be digitally displayed. Rather more expensive instruments will automatically read the vertical angle and directly display the horizontal distance as well as the slope distance. This is very useful when employing what is termed the *stake-out facility* i.e. an indicator of how much movement is required to set the prism at a specified distance when it is being moved on line.

b) Setting out lengths
To allow for *slope correction* when setting-out, the procedure is:

 i) Set out a temporary peg, with or without slope correction.
 ii) Measure the distance accurately with slope correction (make any necessary corrections for ambient conditions, etc.)
 iii) Subtract the corrected value from the required length.
 iv) Tape this difference from the temporary peg to locate the position of the final peg (take care to stay 'on-line').
 v) Check distance to final peg, with slope correction.

This is generally a better method than using a pole-mounted prism and gives better accuracy.

10.1.4 Sources of error affecting accuracy of measurement

Although there is not much that can go wrong with EDM equipment there are some problems which will give unacceptable results.

a) *Weak or non-existent return signal* can be the result of
 i) *Exhausted batteries.* Nothing is worse than to have to stop work because of a lack of power and it is better to carry two batteries at all times fully charged. It is also worthy of note that *nickel-cadmium* batteries should not be recharged until fully discharged, or very nearly so, which necessitates keeping a check of cumulative use. Any residual charge must be discharged according to the manufacturer's instructions prior to recharging, which is usually done overnight.
 ii) *Faulty cable connections.* On older instruments, particularly, where the battery is connected by cables to the 'aiming-head', a regular check of the cable (and the connections) should be made. Carry a spare cable (or set of cables) when working in the field.
b) *Outgoing signal too strong.* Here the risk is of the signal being returned by surfaces other than the prism, which will result in 'short measurement'.
c) *Prism errors.* These will occur if:

 i) *Prisms are not aligned* i.e. not pointed approximately towards the EDM equipment.

ii) *Prisms not centred* which will give inaccurate results. Since there is likely to be a backwards movement by most telescope mounted EDMs when an angle of elevation is set, non-matching prisms will not then be accurately centred. To offset this either a tilting target prism assembly, carefully aligned, should be used or the theodolite and EDM should in turn be pointed towards the prism. This latter will give a negligible error on all but the shortest of sights. Additionally, a tilting prism used with a standard-mounted or an integral EDM system may result in a 'shortening' error to the required distance (see Fig. 10.4).

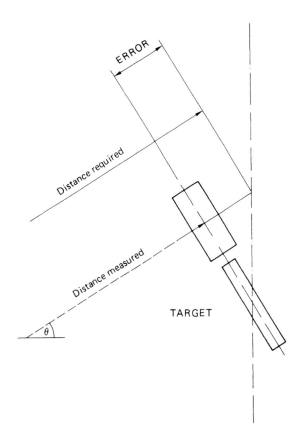

Fig. 10.4 Error induced by tilting prism when used with a standard-mounted or an integral EDM system

iii) *Insufficient number of prisms* used for long distances will result in a poor signal.

d) *Collimation error* occurs when 'add-on' systems are used and the lines of sight (collimation) of the EDM and the theodolite are not parallel. The adjustment is generally simple:
  i) line up the axes optically on a prism and target, then
  ii) adjust the EDM axis to give the maximum return signal.

For large errors, an initial adjustment will have to be carried out by sighting a nearby prism, then moving it progressively away. (A single prism at 300m will usually allow the EDM axis to be satisfactorily aligned).

Collimation adjustment requires to be checked on all new instruments with telescope-mounted apparatus needing to be properly aligned both horizontally and vertically and standard-mounted equipment horizontally only.

e) *Faulty EDM instrument.* When in regular use, the EDM should be checked *at least once each week* against a known length, e.g. a base line set out over acceptably level ground by taping, with all corrections applied. Generally, checking should be done by experienced operators or, in some cases, by the manufacturer. *When the instrument is not in regular use, it should always be checked a few days before work commences.*

f) *Poor visibility* may require 'pointing' at the target to be carried out electronically, the operator turning the horizontal and vertical tangent screws until a maximum return signal is received on the indicator of the instrument.

10.2 Computer facilities in surveying

Computers have several characteristics which make them extremely useful in surveying operations. Broadly these are

a) the great speed at which they can carry out arithmetical operations, enabling the most complex survey calculations to be completed, literally, within seconds;
b) accuracy of calculation, particularly in relation to repetitive tasks;
c) programs can be written for specific survey calculations and, once proved, are there for 'all time'.

The potential for saving of time is not, however, limited to the more complex calculations. Relatively simple operations, such as the calculation of co-ordinates from tacheometric observations, can be speeded up considerably and the saving of time and effort is not just embodied in the speed of the calculation but in the elimination of human error from the calculation process. In addition to the rapid processing of data, computers can store and retrieve data quickly using magnetic discs or tapes and this combination of rapid data storage and retrieval, coupled with equally rapid data processing, allows large amounts of data to be handled efficiently.

Computers have been used for survey purposes from a very early stage in their development. Originally, large mainframe computers were found to be suitable for large, complex calculations, particularly for the adjustment of large control survey networks, and, although a mainframe computer may still be used for exceptionally large or complex tasks, the micro-computer can deal with most work required in the surveyor's office. In the field, the introduction of data loggers has enabled electronically measured observations to be recorded and automatically processed, thus eliminating laborious manual field and calculation work. As a result, a greater number of checks in the field and a better analysis of results can be undertaken in less time than in the past.

10.2.1 Applications

The increasing use of computers in surveying has affected some of the more frequently used methods of surveying as well as creating new techniques.

   a) Resection *(see Section 9.6.4) the operation by which the point to be located is the station occupied by the instrument.* The problem has been described and two methods of solution offered using a plane table. However, the use of a theodolite/EDM to make observations and a computer to make calculations enables more use to be made of a method which has often been used less frequently in the past than its value would merit. Calculations are (or were) fairly long, including the use of several trigonometric functions, and not convenient when using mechanical or electronic calculators, since the reduction process can be tedious and prone to errors.

   However, *resection* is a very useful method for establishing control in a rapidly changing environment such as large civil engineering, quarrying or open-cast mining sites. Fig. 10.5 shows a situation where the observation of five rays (four angles) provides the opportunity to calculate the co-ordinates of the resected point from ten different combinations of three rays. By using a computer, it becomes practical to calculate every possible combination of three observed rays to produce resected co-ordinates, a task often impractical when only simple calculators are available.

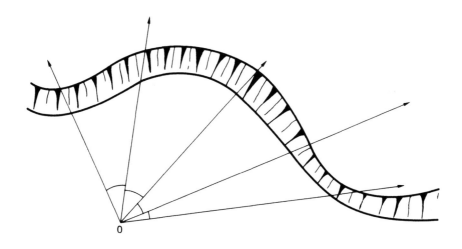

0 = resected point

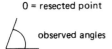
   observed angles

Fig. 10.5 Resection

A further advantage of resection as a method of establishing control is that it is very efficient in terms of fieldwork compared with other methods such as the traverse or triangulation. Computers, then, have implications not only for calculation but also for fieldwork.

b) **Adjustments** Another group of surveying techniques which has been affected by the availability and use of computers is adjustments, for example,

   i) the adjustments made to field observations which can now be made in the field, either manually or automatically, as they are taken, and
   ii) the adjustments applied to calculated co-ordinates to achieve geometric consistency.

Some of these adjustments involve long or very complex calculations, for example, *least squares adjustments of control networks* or *Bowditch's adjustment of a traverse*. These types of adjustment have usually been discarded in favour of simple approximations (see Section 9.6.4 for graphic version of Bowditch), or even ignored completely because of the time-consuming calculations involved. Computers have allowed adjustments to be applied far more easily and, therefore, more widely than had previously been possible and programs for the resolution of traverse survey data, etc., are in general use.

c) **Areas and volumes** Section 8.10 dealt with the calculation of areas and volumes by various methods which, whilst being fairly straightforward, often involve the processing of large amounts of data. Manual or mechanical methods of calculation can be, to some degree, lacking in accuracy depending on the care taken both in the collection of information and its processing as well as tedious when, in order to avoid large errors, time-consuming checks are carried out.

However, using a computer for the calculation of volumes excavated, either in total or periodic part, is not only easily carried out but also can be done at any time thanks to the database/storage facility of the equipment. This, in turn, means that only a relatively small amount of data, i.e. the details of the changes to the volume being measured, need be input before each calculation is made. Volumetric calculation certainly demonstrates the flexibility which the computer brings to the processing of data. For example, volumes of earthworks are frequently calculated from a series of co-ordinated sectional areas set at intervals through the quantity to be measured and, as a check, a second series of sections perpendicular to the first is usually compiled from which the volume is re-calculated. Again, when done manually or mechanically, this is an extremely time-consuming task for all but the smallest volumes. The computer enables data to be quickly and easily arranged to compile sections in any direction or at any interval through the volume being measured so that, in addition to the saving of time, a more flexible use of data is facilitated.

d) **3-D modelling and graphics** Computers can be used to process topographical survey data and produce three-dimensional 'models' of land surfaces. The data is normally entered into the system in the form of spot levels taken at regular intervals over the surface to be 'modelled'. The computer program then produces *a net diagram, i.e. a three-dimensional view of the surface* using either a graphics screen or a plotter. Once the data has been entered the program will allow the surface to be drawn from any viewpoint. It is fairly easy to extend the program to a *computer-aided manufacturing (CAM)* system in order to produce three-dimensional models in timber or plastics, etc., whilst two and three-dimensional drawings may be produced for a model-maker to build a scale-model of the land surface.

These models, whether in drawn or solid form, have many applications, being particularly useful as design aids. For example, it is possible to examine the suitability of alternative landfill surfaces during the restoration works, required by the planning process, following quarrying, open-case mining or tipping operations, as well as siting, layout and design of buildings, roads and other artefacts.

In addition to the production of net diagrams and 'solid' graphics, computers can also be used to interpolate contours from surface spot-levels and vice-versa. When this type of graphic facility is combined with the rapid calculation of volumetric quantities in order, for example, to show the changing landscape due to excavation operations, it is easy to appreciate how the computer system has become a very necessary and powerful tool in surveying practice.

Although the applications discussed in this section are not exhaustive, they serve to illustrate the benefits of the use of computers in the field of surveying. Again, like theodolites, computers do not growl or byte (forgive the pun!) and can become an indispensable part of one's working life providing a proper grounding and understanding is acquired from the outset.

10.2.2 Summary

Initially, computers were used principally as sophisticated calculators to facilitate the rapid execution of previously tedious, repetitive and time-consuming tasks of calculation. However, computers are now being used for an increasing range of surveying applications, the benefits of which have resulted in several important implications,

i) the selection of field methods for particular tasks need no longer be conditional on computational simplicity, since the most complex survey calculations can be executed extremely rapidly;

ii) the use of survey methods can be tailored more closely to the requirements of the task in prospect and other conditions which need to be taken into account;

iii) the surveyor can make a more discerning and flexible use of the methods available to him;

iv) the saving of time and effort gained in both field and office work, resulting from the use of the computer, may be applied to other priorities;

v) by taking over many of the more mundane and time-consuming tasks, computers enable the surveyor to concentrate time and effort on the professional rather than the mechanical aspects of surveying practice.

Even with the rapid developments in technology of recent years, much of the huge potential of computers for survey work is still untapped by the majority of users. Concepts such as the *integrated survey system* (see Section

10.3), although more and more widely used, are still seen as expensive items suitable only for major projects. The computer certainly has a role to play in the future of surveying and the expansion of this role as the twentieth century draws to a close is expected to be rapid.

10.3 Integrated Survey Systems

In some situations, for example, large, progressive excavation and mining operations, the landscape changes rapidly due to the sheer bulk excavation capacities of the machinery being used and the deadlines to be met which can require the work to be carried out at a fast rate. Often a new drawing showing levels and other information (i.e. daily and cumulative volumetric totals of material excavated) is required at the start of each day's work. It is obvious that this cannot be done effectively by 'conventional' or 'traditional' methods and a combination of electronic surveying techniques and computer facilities is essential if delay is to be avoided. For many years it has been possible to carry out the following procedure:

i) at the end of the day's work, take observations in the field relative to distance and height measurement using EDM equipment with automatic tracking and reading facilities;

ii) analyse the information gathered using 'add-on' equipment (see Fig. 10.6) or, latterly, integral computing facilities;

iii) transmit the data/information *via telephone lines to an office database* – often miles from the site;

iv) introduce the data into a program for use by a *computer-aided draughting (CAD) system*;

v) produce, overnight, new drawn information which can be delivered to site ready for the start of a new day's work.

From earlier parts of the book, it can be seen that the basis of any survey system consists of three elements,

a) data collection – the input into the system,

b) data processing, and

c) data presentation – the output of the system.

A fully integrated system encompasses everything from the field observations to the production of the final drawings and each of the elements must be linked to provide a continuity of operation. Once data has entered the system it remains within it until it emerges in its final form as a plan, cross-section, 3-D drawing and so on. It is also fairly obvious that each of the three elements must be compatible with the other two in terms of hardware and software if data is to pass between them.

The *interfacing* of the differing elements of a system has developed rapidly over the last decade and standard interface hardware is now available for most computers, plotters and electronic surveying instruments.

a) **Total Station** a good definition of which is *a combination of theodolite, EDM equipment and data processor which gives x, y and z co-ordinates directly*. There is, however, some disagreement over the term 'Total Station', in that a combined instrument with a *data logging socket* is known as an *electronic* *tacheometer* which would suggest that to be a 'total station' the instrument should be a *precise distancer/theodolite* with a considerable range of integral computing capability.

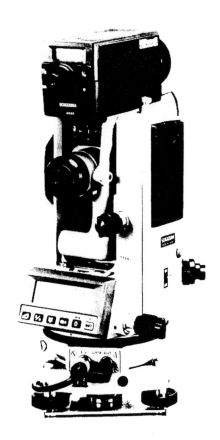

Fig. 10.6 'Sokkisha' theodolite with 'add-on' EDM above telescope

It is also possible to add an *EDM unit* to an *electronic theodolite with calculation and data output facilities* in the form of a microchip-based processing and storage unit. When the instrument is operated, the values for distance, horizontal and vertical angle are automatically recorded and in some cases the instrument can process these values to produce co-ordinates. The quantities measured and/or the calculated values can either be stored in an internal memory or on a magnetic medium such as a cassette tape and retrieved later. This is the data collection function of the survey system.

When used as part of an integrated system, the 'Total Station' instrument (see Fig. 10.7) has two advantages over rather more conventional pre-decessors,

    i) instrument reading errors are eliminated from field observations by the automatic recording of the measured values, and

ii) human data handling is eliminated by recording the measured values in a form which can be directly input to a computer for processing.

Automatic operation speeds up both the initial data recording and the data transfer processes.

(a)   (b)

Fig. 10.7 'Total Station' instrument: (a) 'Sokkisha' electronic tacheometer, (b) 'Sokkisha' data logger (electronic field-book)

b) **Field data loggers** which, in appearance, look like a cross between a calculator and a micro-computer, may be instrument-mounted, tripod-fixed or hand-held. The logger is connected to the theodolite/EDM by a cable and power is provided by a re-chargeable nickel-cadmium battery (see Section 10.1.4) and has sufficient memory space for the storage of a day's work. Some loggers have *interchangeable memory modules*, whilst, in the office, loggers can be connected by a special cable to microcomputers. There are two types

   i) *a general purpose logger* which requires programming for manual and automatic recording. Programming, so that data can be automatically recorded, transferred to an office micro or printed or plotted directly, must be done to suit the instruments with which the logger is being used.
   ii) *a dedicated logger* which is produced to provide automatic recording when interfaced with the appropriate field instruments, e.g. electronic theodolites and EDM, and pre-programmed for recording and transfer of data.

All field data, whether surveyed electronically or not, can be stored by computer to form what is termed *a digital ground model (DGM)*. Descriptive information and co-ordinates in x, y *and* z directions for each point surveyed are noted. It is more practical to purchase the necessary software, which can be tailored to the user's needs, when buying data recording equipment than it is to devise one's own program for logging, transfer, retrieval, calculations, storage and plotting.

c) **Office microcomputers** with adequate memory space, monitor, disk drive and printer can be programmed to receive data from a logger, to store data, to perform calculations and to drive a plotter where automated plotting is a requirement. Simple programs for the production of plans in several colours, with points connected by straight lines, are commercially available, as are more complex programs, as discussed in Section 10.3.

d) **Data processing software** can be extremely sophisticated, although there is still a need for human interaction with the system to enable amendments to the input data to be made. In a fully integrated system, processing of data requires a large computer processing capacity, since the basic calculations represent only a small fraction of the processing required. The remainder of the processing capacity represents the difference between the production of a simple list of co-ordinates and a well drawn, fully annotated, plan.

Although many graphical operations in map/plan production are apparently straightforward to the human brain, a computer often requires quite complex software packages to achieve the same results. Sophisticated *computer graphics packages* are now readily available and, in terms of cost, are within the range of the small user.

Additionally, it is also possible to purchase *Ordnance Survey map data* for most areas of the UK in digital form on magnetic tape or disk. This data can be plotted on the user's own equipment to produce maps at scales limited only to the capability of the plotter and is particularly useful in regard to control survey work and urban design work.

e) **Output** can be in the form of plans, sections, net diagrams, tables (using *spreadsheet techniques*) and so on, which may be produced on *plotters* and *printers* controlled by the computer (See Fig. 10.8). The quality of the results will vary with the printing system which may range from *daisywheel* through *dot-matrix* to the latest *laser* techniques. Output may be in 'black and white' or colour and some of the results from the more sophisticated (and expensive) graphics systems are on a par with the artwork of the book world.

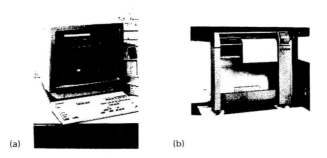

(a)   (b)

Fig. 10.8 Plotting by computer: (a) Typical VDU showing 'Sokkisha' SDR map software (keyboard standard unit), (b) Typical plotter, (c) Plotted drawing

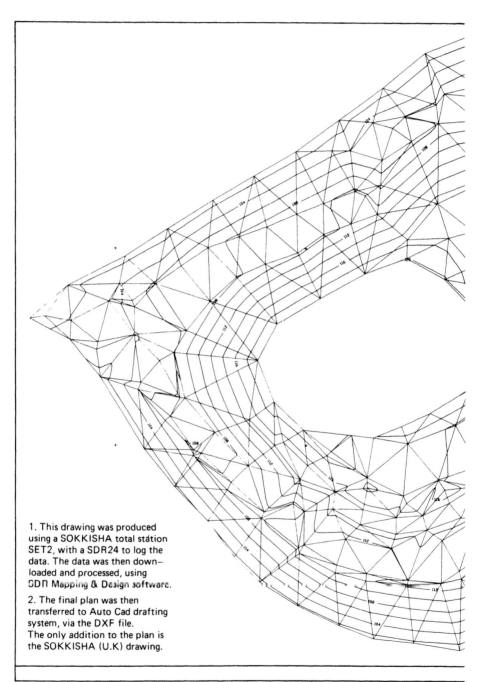

1. This drawing was produced using a SOKKISHA total státion SET2, with a SDR24 to log the data. The data was then down—loaded and processed, using SDR Mapping & Design software.

2. The final plan was then transferred to Auto Cad drafting system, via the DXF file.
The only addition to the plan is the SOKKISHA (U.K) drawing.

(c)

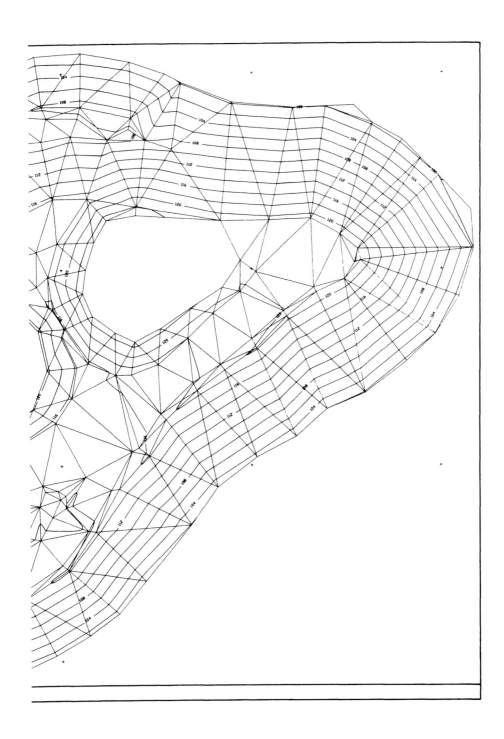

Pre-selected parts or areas of a diagram or map may be plotted to a larger scale, in order to show detail, by using the '*zoom*' facility built into the program. This flexibility is a major advantage where the survey data has many users, each with different requirements.

Although each of the groups of equipment making up an 'integrated survey system' can be used in a stand-alone context, the benefits which they produce when used in this way are generally far more limited than when they are used as part of a fully integrated system.

10.4 Lasers

Lasers are setting-out instruments which are used as an alternative to the engineer's level and theodolite for a range of tasks on site. *A narrow visible beam of red light* is produced by *helium-neon(He-Ne) gas* whilst *an invisible beam* is produced by *gallium-arsenide*. The instrument provides *an optical line if fixed as in an alignment laser* or *an optical plane if rotated, as in a rotating laser*. Lasers are defined by BS 4803 as Class 1, 2, 3A and 3B in an ascending order of power. Fig 10.9 illustrates the types of lasers and detectors in general use.

Fig. 10.9 Setting-out equipment. Left: 'Sokkisha' level planer (main unit). Right: 'Sokkisha' photoelectric beam detector

10.4.1 Applications

Lasers are used extensively in building construction, pipe-laying and tunnelling work with site instruments at the lower-powered end of the instrument range.

A visible-beam is the most useful and convenient light when using an alignment laser whilst both the visible-beam and invisible-beam are used for the rotating laser, with the latter perhaps favoured to avoid the nuisance of a visible *occulting beam*. The beam can be seen where it meets an obstruction and 'lining-in' and 'boning' can be carried out using the typical template or traveller of appropriate dimensions. Instead of the latter, a *photoelectric detector* may be used at the position to be aligned (see Fig. 10.9(b)).

a) **an alignment laser with a visible beam** is used as follows:

  i) *for pipe-laying,* a laser is set to line and gradient
  • for *the excavation* of the trench, using a traveller with a solid target;
  • for *the laying* of the pipes, using a transparent target inside the pipe(s).

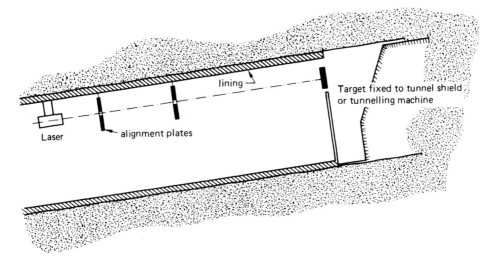

Fig. 10.10 Tunnelling procedure

  ii) *for tunnelling* the normal procedure (see Fig. 10.10) is that

  • two plates are fixed to the completed tunnel roof such that small holes in each plate define a line offset from the centre-line of the tunnel. (The centre of the tunnel is defined by standard setting-out techniques.);
  • a laser is fixed to the roof of the tunnel and adjusted until the beam passes through the holes in the two plates;
  • the beam projects on to a target on the tunnel shield or tunnelling machine for control of line and gradient;
  • the laser is moved forward and reset when the target distance reaches maximum (for accuracy).

b) **a rotating laser** with visible or invisible beam has the following uses

  i) earth-moving to level on the horizontal plane;
  ii) providing a horizontal reference plane across a site;
  iii) providing a vertical reference plane for, as an example, the fixing of curtain walling to a frame construction.

  All rotating lasers should be *gyro-compensated* and able to operate in both the horizontal and vertical plane (see Fig. 10.11).

c) **photo-electric detectors,** fixed on a pole, can be used to determine the amount of cut or fill required at strategic points. When being used in alignment work, the detector will indicate the amount of further movement required as the correct alignment position is approached. Detectors may be fitted to excavating and grading machines with a display indicating to the driver whether cut or fill is required.

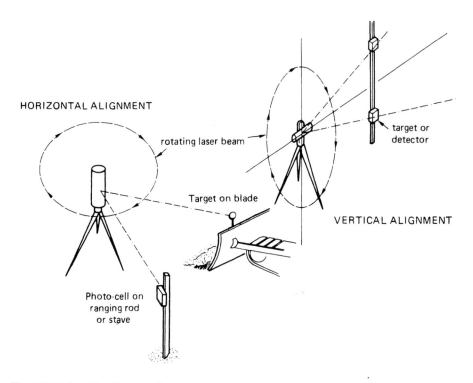

HORIZONTAL ALIGNMENT

rotating laser beam

target or detector

Target on blade

VERTICAL ALIGNMENT

Photo-cell on ranging rod or stave

Fig. 10.11 Rotating laser techniques

10.4.2 Safety

Lasers are a source of danger to operatives if care is not taken. Warnings should be posted on sites and reference made to the BS 4803 'Safety and the Use of Lasers'.

To avoid eye damage from lasers

i) where practicable, use Class 1 or Class 2 lasers which are effectively low-powered,
ii) if Class 3A or 3B lasers must be used, operatives should comply with safety procedures as recommended by the RICS (see also 'Safety in the use of lasers on site', CIOB Technical Information Paper No. 22(1983). Cox E.A.)
iii) operatives should **not**

    a) look along the beam, or
    b) search for it with a level or theodolite by looking through the telescope.

10.4.3 Limitations

Lasers do have some minor limitations which include
  i) *beam divergence* which is typically 5–10mm per 100m.
  ii) *beam refraction* which is caused by

  • non-uniform temperature,
  • beam passing too near to solid material e.g. tunnel wall.

  N.B. When working in tunnels or ducts, they should be ventilated to achieve a uniform temperature.

  iii) *the invisible-beam laser* needing to be used in conjunction with a photo-electric detector.

10.5 Aerial Photography and Photogrammetry

This branch of surveying technique is a combination of *aerial photography, photogrammetry* and *basic ground survey.*

Ever since the first camera was developed, attempts have been made to use photography as an aid to surveying and the first success came when the camera was used as a *photographic plane table* in the survey of high mountain areas. Much commercial and government survey work is now carried out using aerial photography, especially where large areas are being covered, and the procedure is:

  i) photographs are taken from the air, using varying techniques, which replace the detail survey work done on the ground;
  ii) basic survey techniques are used to locate control points, on the ground, in order to give *scale, direction* and *height-control* to the measurements taken from the photographs;
  iii) interpolation of the data takes the form of

  a) plotting accurately on the photograph the positions of the ground control points, and
  b) measuring the detail from the photograph (i.e. photogrammetry).

Visual three-dimensional images can be obtained by the use of *stereoscopic photography, i.e. two pictures of the same area from different viewpoints* which are used in *cartography (map-making).*

a) **Aerial Surveying** is the technique by which the primary observations of the survey are recorded in the form of aerial photographs taken from an aircraft when in level flight.

  Fig. 10.12 shows how *vertical photographs* are taken with the axis of the camera and lens as near vertical as is practical. Other photographs with less importance for the surveyor are produced with the camera axis inclined to the vertical and are termed *oblique photographs.*

  When an area is being photographed for mapping purposes, the aircraft flies in a straight line at a predetermined height, taking successively overlapping photographs at timed intervals. This produces a strip of photographs of a strip of the ground. Adjacent strips are photographed so that they also overlap each other, termed *sidelap*, to ensure that there are no gaps in the work and that the whole of the terrain is covered. Figure 10.13 shows an aerial photograph of Bristol and the interpretation of the features on the ground.

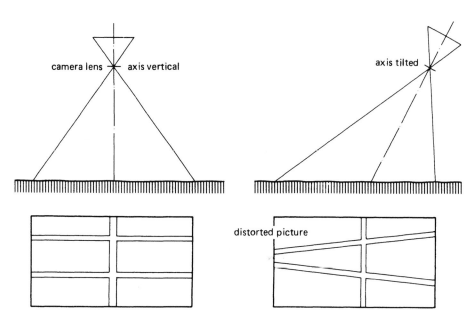

Fig. 10.12 Vertical photography

b) **Photogrammetry** is the technique of taking measurements from photographs. Photogrammetry is dependent upon, and therefore makes use of, the human ability to see in three dimensions, i.e. stereoscopically. By photographing an object from two different camera stations then, by arranging the photographs in such a way that one is seen by the left eye at the same time as the other is seen by the right eye, an impression of solidarity and depth is created. This is known as *a spatial model* and pairs of photographs are viewed through *stereoscopes*. This system was the basis of a highly commercial Edwardian pastime where portraits, buildings, scenic views from home and abroad, etc., were viewed in 3-D in the comfort of the drawingroom.

Applications and practical uses are almost without limit and include the study of both land and buildings (see Section 7.4.7 and Fig. 10.14). It is possible to take measurements on a pair of photographs when viewed stereoscopically, i.e. from the spatial model, which can then be related directly to the ground model.

The techniques of aerial photography and photogrammetry are constantly being developed for an ever wider range of applications. Again, the technique is the subject of more advanced study than the present text will allow.

Fig. 10.13 Aerial photograph and interpretation

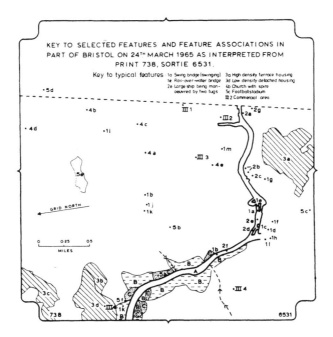

KEY TO SELECTED FEATURES AND FEATURE ASSOCIATIONS IN
PART OF BRISTOL ON 24ᵀᴴ MARCH 1965 AS INTERPRETED FROM
PRINT 738, SORTIE 6531.

Key to typical features:
1a Swing bridge (swinging) 3a High density terrace housing
1e Rail-over-water bridge 3d Low density detached housing
2e Large ship being man- 4b Church with spire
oeuvred by two tugs 5c Football stadium
Ⅲ 2 Commercial area

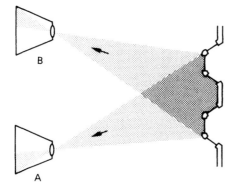

(a) Photographs are taken from
    two different viewpoints

(b) An example of a photogrammetrical
    drawing produced by The Photo-
    grammetric Unit, University of York
    Copyright 'English Heritage'.

Fig. 10.14 Photogrammetry techniques for the reproduction of the elevations of buildings

10.6 Summary

As stated at the outset, this chapter has simply been an introduction to some of the more readily identifiable of modern developments in the field of surveying. The topics touched on, therefore, do not form an exhaustive list and to have dealt with them, or other topics such as *satellite doppler surveys; Mercator projection* and so on, in any great detail would have been against the philosophy of this textbook as a whole.

   It is hoped that for those readers moving to more advanced study, this chapter will help to ease the transition from lower/middle level work to that of a higher order. For those who are not, at least there will be an awareness of some of the more advanced techniques available to the surveyor, some of which, perhaps, will be followed up out of pure interest.

Exercises on Chapter 10
1. List the possible advantages of the use of computers in surveying.
2. Describe an 'Integrated Survey System'.
3. What additional advantages does an 'integrated survey system' have over its conventional predecessors?
4. Describe by means of an annotated sketch what you understand by the term 'Total Station'.

Index